true since the first group of hum
land as "ours" as opposed to "the

The field of border studies

Reflecting their centrality to human social interaction and the exercise of power, borders have become hot research topics across a number of disciplines. Since the late 1960s, geographers, sociologists, anthropologists, economists, environmental psychologists, political scientists, legal scholars, and historians have challenged prior scholarship presuming borders play rather passive roles in international and intrastate relations. Rather than simple pretexts for conflict or impediments to mobility, border scholars began to consider the lines that divide human groups as key processes of multiscalar geopolitics worthy of deeper and more textured consideration. There are many reasons for the relative neglect that border studies endured through much of the twentieth century, but the most important was the assumption of a fixed relationship between power and territory inherent in the nation-state system.

Despite the efforts of the growing border studies community, this assumption has only recently come to be questioned by politicians, the broader academy, and the general public. Few fully appreciate that very different modes of territorial organization existed merely three centuries ago. The sociopolitical order resulting from the transition of monarchical rule by hereditary nobility to democratic governments of elected representatives, along with urbanization, industrialization, and the dissemination of "enlightenment values," is a relatively recent and spatially uneven phenomenon. Nevertheless, these events gave rise to new territorial assumptions and practices that had a profound effect on the way people perceive themselves and their respective places in the world.

The replacement of frontiers with clearly delineated borders reframed human identity and most social processes by ensnaring them in what political geographers commonly refer to as the "territorial trap." This concept derives from three interrelated assumptions. The first is that states are the exclusive arbiters of power within their territories. In other words, states are invested with sovereignty. The second assumption holds that domestic (internal) and foreign (external) affairs are different realms of political and social activity. Therefore, each realm operates with fundamentally different standards of legality and morality. The third assumption views the boundaries of the state as matching the boundaries of the society. In other words, states act as rigid containers that neatly partition global space into nation-state territories corresponding to distinct societies.

Combined with the emotional appeal of nationalism, these assumptions reinforced one another to convey a sense of historical legitimacy and contemporary permanence for a state-centered view of power and its delineation of global space. As economic and social practices became increasingly associated with the state during the nineteenth century, business, labor, politics, sports, military, education, and art came to be viewed through the prism of the territorial trap. The idea that these things were most efficient and pragmatic in relation to the state was advanced and "scientifically" supported through the new social sciences of economics, sociology, and political science. The rise of the modern academy was very much in service to the nation-state. With the church receding as the primary educator of the populace, the framing and shaping of identity to fit the new religion of the nation (i.e., nationalism) fell to governments and their schools and universities. The intense nationalism accompanying World War I, World War II, and the Cold War furthered a general consensus among social scientists that local communities and allegiances would gradually give way to national societies.

In short, throughout much of the twentieth century, the territorial trap relegated research on geographic boundaries to subfields like political anthropology, political geography, or regional politics,

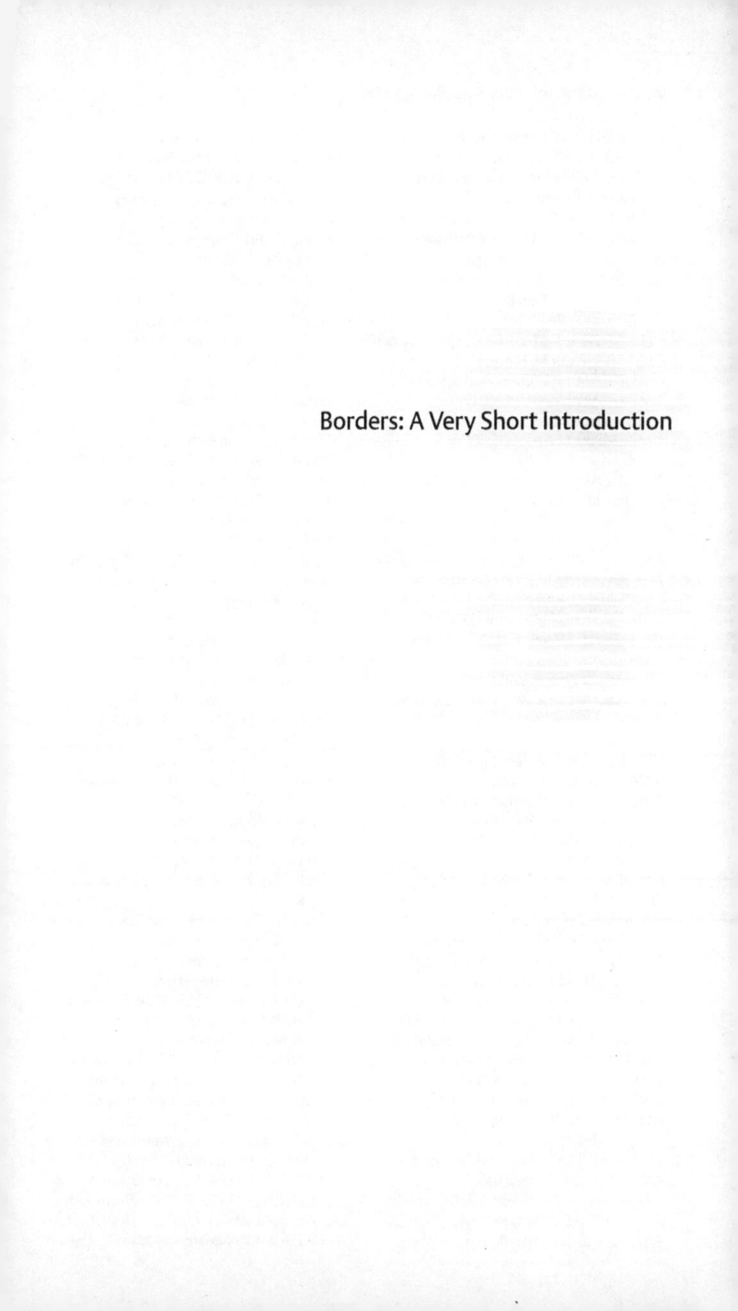

Borders: A Very Short Introduction

Very Short Introductions available now:

ACCOUNTING Christopher Nobes
ADVERTISING Winston Fletcher
AFRICAN AMERICAN RELIGION Eddie S. Glaude Jr
AFRICAN HISTORY John Parker and Richard Rathbone
AFRICAN RELIGIONS Jacob K. Olupona
AGNOSTICISM Robin Le Poidevin
AGRICULTURE Paul Brassley and Richard Soffe
ALEXANDER THE GREAT Hugh Bowden
ALGEBRA Peter M. Higgins
AMERICAN HISTORY Paul S. Boyer
AMERICAN IMMIGRATION David A. Gerber
AMERICAN LEGAL HISTORY G. Edward White
AMERICAN POLITICAL HISTORY Donald Critchlow
AMERICAN POLITICAL PARTIES AND ELECTIONS L. Sandy Maisel
AMERICAN POLITICS Richard M. Valelly
THE AMERICAN PRESIDENCY Charles O. Jones
THE AMERICAN REVOLUTION Robert J. Allison
AMERICAN SLAVERY Heather Andrea Williams
THE AMERICAN WEST Stephen Aron
AMERICAN WOMEN'S HISTORY Susan Ware
ANAESTHESIA Aidan O'Donnell
ANARCHISM Colin Ward
ANCIENT ASSYRIA Karen Radner
ANCIENT EGYPT Ian Shaw
ANCIENT EGYPTIAN ART AND ARCHITECTURE Christina Riggs
ANCIENT GREECE Paul Cartledge
THE ANCIENT NEAR EAST Amanda H. Podany
ANCIENT PHILOSOPHY Julia Annas
ANCIENT WARFARE Harry Sidebottom
ANGELS David Albert Jones
ANGLICANISM Mark Chapman
THE ANGLO-SAXON AGE John Blair
THE ANIMAL KINGDOM Peter Holland
ANIMAL RIGHTS David DeGrazia
THE ANTARCTIC Klaus Dodds
ANTISEMITISM Steven Beller
ANXIETY Daniel Freeman and Jason Freeman
THE APOCRYPHAL GOSPELS Paul Foster
ARCHAEOLOGY Paul Bahn
ARCHITECTURE Andrew Ballantyne
ARISTOCRACY William Doyle
ARISTOTLE Jonathan Barnes
ART HISTORY Dana Arnold
ART THEORY Cynthia Freeland
ASTROBIOLOGY David C. Catling
ASTROPHYSICS James Binney
ATHEISM Julian Baggini
AUGUSTINE Henry Chadwick
AUSTRALIA Kenneth Morgan
AUTISM Uta Frith
THE AVANT GARDE David Cottington
THE AZTECS Davíd Carrasco
BACTERIA Sebastian G. B. Amyes
BARTHES Jonathan Culler
THE BEATS David Sterritt
BEAUTY Roger Scruton
BESTSELLERS John Sutherland
THE BIBLE John Riches
BIBLICAL ARCHAEOLOGY Eric H. Cline
BIOGRAPHY Hermione Lee
BLACK HOLES Katherine Blundell
THE BLUES Elijah Wald
THE BODY Chris Shilling
THE BOOK OF MORMON Terryl Givens
BORDERS Alexander C. Diener and Joshua Hagen
THE BRAIN Michael O'Shea
BRICS Andrew F. Cooper
THE BRITISH CONSTITUTION Martin Loughlin
THE BRITISH EMPIRE Ashley Jackson
BRITISH POLITICS Anthony Wright
BUDDHA Michael Carrithers
BUDDHISM Damien Keown
BUDDHIST ETHICS Damien Keown
BYZANTIUM Peter Sarris
CANCER Nicholas James
CAPITALISM James Fulcher
CATHOLICISM Gerald O'Collins
CAUSATION Stephen Mumford and Rani Lill Anjum
THE CELL Terence Allen and Graham Cowling
THE CELTS Barry Cunliffe
CHAOS Leonard Smith
CHEMISTRY Peter Atkins
CHILD PSYCHOLOGY Usha Goswami
CHILDREN'S LITERATURE Kimberley Reynolds
CHINESE LITERATURE Sabina Knight
CHOICE THEORY Michael Allingham
CHRISTIAN ART Beth Williamson
CHRISTIAN ETHICS D. Stephen Long
CHRISTIANITY Linda Woodhead
CITIZENSHIP Richard Bellamy
CIVIL ENGINEERING David Muir Wood
CLASSICAL LITERATURE William Allan
CLASSICAL MYTHOLOGY Helen Morales
CLASSICS Mary Beard and John Henderson

CLAUSEWITZ Michael Howard
CLIMATE Mark Maslin
CLIMATE CHANGE Mark Maslin
THE COLD WAR Robert McMahon
COLONIAL AMERICA Alan Taylor
COLONIAL LATIN AMERICAN LITERATURE Rolena Adorno
COMBINATORICS Robin Wilson
COMEDY Matthew Bevis
COMMUNISM Leslie Holmes
COMPLEXITY John H. Holland
THE COMPUTER Darrel Ince
COMPUTER SCIENCE Subrata Dasgupta
CONFUCIANISM Daniel K. Gardner
THE CONQUISTADORS Matthew Restall and Felipe Fernández-Armesto
CONSCIENCE Paul Strohm
CONSCIOUSNESS Susan Blackmore
CONTEMPORARY ART Julian Stallabrass
CONTEMPORARY FICTION Robert Eaglestone
CONTINENTAL PHILOSOPHY Simon Critchley
CORAL REEFS Charles Sheppard
CORPORATE SOCIAL RESPONSIBILITY Jeremy Moon
CORRUPTION Leslie Holmes
COSMOLOGY Peter Coles
CRIME FICTION Richard Bradford
CRIMINAL JUSTICE Julian V. Roberts
CRITICAL THEORY Stephen Eric Bronner
THE CRUSADES Christopher Tyerman
CRYPTOGRAPHY Fred Piper and Sean Murphy
CRYSTALLOGRAPHY A. M. Glazer
THE CULTURAL REVOLUTION Richard Curt Kraus
DADA AND SURREALISM David Hopkins
DANTE Peter Hainsworth and David Robey
DARWIN Jonathan Howard
THE DEAD SEA SCROLLS Timothy Lim
DEMOCRACY Bernard Crick
DERRIDA Simon Glendinning
DESCARTES Tom Sorell
DESERTS Nick Middleton
DESIGN John Heskett
DEVELOPMENTAL BIOLOGY Lewis Wolpert
THE DEVIL Darren Oldridge
DIASPORA Kevin Kenny
DICTIONARIES Lynda Mugglestone
DINOSAURS David Norman
DIPLOMACY Joseph M. Siracusa
DOCUMENTARY FILM Patricia Aufderheide
DREAMING J. Allan Hobson
DRUGS Leslie Iversen
DRUIDS Barry Cunliffe
EARLY MUSIC Thomas Forrest Kelly
THE EARTH Martin Redfern
EARTH SYSTEM SCIENCE Tim Lenton
ECONOMICS Partha Dasgupta
EDUCATION Gary Thomas
EGYPTIAN MYTH Geraldine Pinch
EIGHTEENTH-CENTURY BRITAIN Paul Langford
THE ELEMENTS Philip Ball
EMOTION Dylan Evans
EMPIRE Stephen Howe
ENGELS Terrell Carver
ENGINEERING David Blockley
ENGLISH LITERATURE Jonathan Bate
THE ENLIGHTENMENT John Robertson
ENTREPRENEURSHIP Paul Westhead and Mike Wright
ENVIRONMENTAL ECONOMICS Stephen Smith
ENVIRONMENTAL POLITICS Andrew Dobson
EPICUREANISM Catherine Wilson
EPIDEMIOLOGY Rodolfo Saracci
ETHICS Simon Blackburn
ETHNOMUSICOLOGY Timothy Rice
THE ETRUSCANS Christopher Smith
THE EUROPEAN UNION John Pinder and Simon Usherwood
EVOLUTION Brian and Deborah Charlesworth
EXISTENTIALISM Thomas Flynn
EXPLORATION Stewart A. Weaver
THE EYE Michael Land
FAMILY LAW Jonathan Herring
FASCISM Kevin Passmore
FASHION Rebecca Arnold
FEMINISM Margaret Walters
FILM Michael Wood
FILM MUSIC Kathryn Kalinak
THE FIRST WORLD WAR Michael Howard
FOLK MUSIC Mark Slobin
FOOD John Krebs
FORENSIC PSYCHOLOGY David Canter
FORENSIC SCIENCE Jim Fraser
FORESTS Jaboury Ghazoul
FOSSILS Keith Thomson
FOUCAULT Gary Gutting
THE FOUNDING FATHERS R. B. Bernstein
FRACTALS Kenneth Falconer
FREE SPEECH Nigel Warburton
FREE WILL Thomas Pink
FRENCH LITERATURE John D. Lyons
THE FRENCH REVOLUTION William Doyle

FREUD Anthony Storr
FUNDAMENTALISM Malise Ruthven
FUNGI Nicholas P. Money
GALAXIES John Gribbin
GALILEO Stillman Drake
GAME THEORY Ken Binmore
GANDHI Bhikhu Parekh
GENES Jonathan Slack
GENIUS Andrew Robinson
GEOGRAPHY John Matthews and David Herbert
GEOPOLITICS Klaus Dodds
GERMAN LITERATURE Nicholas Boyle
GERMAN PHILOSOPHY Andrew Bowie
GLOBAL CATASTROPHES Bill McGuire
GLOBAL ECONOMIC HISTORY Robert C. Allen
GLOBALIZATION Manfred Steger
GOD John Bowker
GOETHE Ritchie Robertson
THE GOTHIC Nick Groom
GOVERNANCE Mark Bevir
THE GREAT DEPRESSION AND THE NEW DEAL Eric Rauchway
HABERMAS James Gordon Finlayson
HAPPINESS Daniel M. Haybron
HEGEL Peter Singer
HEIDEGGER Michael Inwood
HERMENEUTICS Jens Zimmermann
HERODOTUS Jennifer T. Roberts
HIEROGLYPHS Penelope Wilson
HINDUISM Kim Knott
HISTORY John H. Arnold
THE HISTORY OF ASTRONOMY Michael Hoskin
THE HISTORY OF CHEMISTRY William H. Brock
THE HISTORY OF LIFE Michael Benton
THE HISTORY OF MATHEMATICS Jacqueline Stedall
THE HISTORY OF MEDICINE William Bynum
THE HISTORY OF TIME Leofranc Holford-Strevens
HIV/AIDS Alan Whiteside
HOBBES Richard Tuck
HOLLYWOOD Peter Decherney
HORMONES Martin Luck
HUMAN ANATOMY Leslie Klenerman
HUMAN EVOLUTION Bernard Wood
HUMAN RIGHTS Andrew Clapham
HUMANISM Stephen Law
HUME A. J. Ayer
HUMOUR Noël Carroll
THE ICE AGE Jamie Woodward
IDEOLOGY Michael Freeden
INDIAN PHILOSOPHY Sue Hamilton
INFECTIOUS DISEASE Marta L. Wayne and Benjamin M. Bolker
INFORMATION Luciano Floridi
INNOVATION Mark Dodgson and David Gann
INTELLIGENCE Ian J. Deary
INTERNATIONAL LAW Vaughan Lowe
INTERNATIONAL MIGRATION Khalid Koser
INTERNATIONAL RELATIONS Paul Wilkinson
INTERNATIONAL SECURITY Christopher S. Browning
IRAN Ali M. Ansari
ISLAM Malise Ruthven
ISLAMIC HISTORY Adam Silverstein
ITALIAN LITERATURE Peter Hainsworth and David Robey
JESUS Richard Bauckham
JOURNALISM Ian Hargreaves
JUDAISM Norman Solomon
JUNG Anthony Stevens
KABBALAH Joseph Dan
KAFKA Ritchie Robertson
KANT Roger Scruton
KEYNES Robert Skidelsky
KIERKEGAARD Patrick Gardiner
KNOWLEDGE Jennifer Nagel
THE KORAN Michael Cook
LANDSCAPE ARCHITECTURE Ian H. Thompson
LANDSCAPES AND GEOMORPHOLOGY Andrew Goudie and Heather Viles
LANGUAGES Stephen R. Anderson
LATE ANTIQUITY Gillian Clark
LAW Raymond Wacks
THE LAWS OF THERMODYNAMICS Peter Atkins
LEADERSHIP Keith Grint
LIBERALISM Michael Freeden
LIGHT Ian Walmsley
LINCOLN Allen C. Guelzo
LINGUISTICS Peter Matthews
LITERARY THEORY Jonathan Culler
LOCKE John Dunn
LOGIC Graham Priest
LOVE Ronald de Sousa
MACHIAVELLI Quentin Skinner
MADNESS Andrew Scull
MAGIC Owen Davies
MAGNA CARTA Nicholas Vincent
MAGNETISM Stephen Blundell
MALTHUS Donald Winch
MANAGEMENT John Hendry
MAO Delia Davin
MARINE BIOLOGY Philip V. Mladenov

THE MARQUIS DE SADE John Phillips
MARTIN LUTHER Scott H. Hendrix
MARTYRDOM Jolyon Mitchell
MARX Peter Singer
MATERIALS Christopher Hall
MATHEMATICS Timothy Gowers
THE MEANING OF LIFE Terry Eagleton
MEDICAL ETHICS Tony Hope
MEDICAL LAW Charles Foster
MEDIEVAL BRITAIN John Gillingham and Ralph A. Griffiths
MEDIEVAL LITERATURE Elaine Treharne
MEDIEVAL PHILOSOPHY John Marenbon
MEMORY Jonathan K. Foster
METAPHYSICS Stephen Mumford
THE MEXICAN REVOLUTION Alan Knight
MICHAEL FARADAY Frank A. J. L. James
MICROBIOLOGY Nicholas P. Money
MICROECONOMICS Avinash Dixit
MICROSCOPY Terence Allen
THE MIDDLE AGES Miri Rubin
MINERALS David Vaughan
MODERN ART David Cottington
MODERN CHINA Rana Mitter
MODERN DRAMA Kirsten E. Shepherd-Barr
MODERN FRANCE Vanessa R. Schwartz
MODERN IRELAND Senia Pašeta
MODERN JAPAN Christopher Goto-Jones
MODERN LATIN AMERICAN LITERATURE Roberto González Echevarría
MODERN WAR Richard English
MODERNISM Christopher Butler
MOLECULES Philip Ball
THE MONGOLS Morris Rossabi
MOONS David A. Rothery
MORMONISM Richard Lyman Bushman
MOUNTAINS Martin F. Price
MUHAMMAD Jonathan A. C. Brown
MULTICULTURALISM Ali Rattansi
MUSIC Nicholas Cook
MYTH Robert A. Segal
THE NAPOLEONIC WARS Mike Rapport
NATIONALISM Steven Grosby
NELSON MANDELA Elleke Boehmer
NEOLIBERALISM Manfred Steger and Ravi Roy
NETWORKS Guido Caldarelli and Michele Catanzaro
THE NEW TESTAMENT Luke Timothy Johnson
THE NEW TESTAMENT AS LITERATURE Kyle Keefer
NEWTON Robert Iliffe
NIETZSCHE Michael Tanner
NINETEENTH-CENTURY BRITAIN Christopher Harvie and H. C. G. Matthew
THE NORMAN CONQUEST George Garnett
NORTH AMERICAN INDIANS Theda Perdue and Michael D. Green
NORTHERN IRELAND Marc Mulholland
NOTHING Frank Close
NUCLEAR PHYSICS Frank Close
NUCLEAR POWER Maxwell Irvine
NUCLEAR WEAPONS Joseph M. Siracusa
NUMBERS Peter M. Higgins
NUTRITION David A. Bender
OBJECTIVITY Stephen Gaukroger
THE OLD TESTAMENT Michael D. Coogan
THE ORCHESTRA D. Kern Holoman
ORGANIZATIONS Mary Jo Hatch
PAGANISM Owen Davies
THE PALESTINIAN-ISRAELI CONFLICT Martin Bunton
PARTICLE PHYSICS Frank Close
PAUL E. P. Sanders
PEACE Oliver P. Richmond
PENTECOSTALISM William K. Kay
THE PERIODIC TABLE Eric R. Scerri
PHILOSOPHY Edward Craig
PHILOSOPHY IN THE ISLAMIC WORLD Peter Adamson
PHILOSOPHY OF LAW Raymond Wacks
PHILOSOPHY OF SCIENCE Samir Okasha
PHOTOGRAPHY Steve Edwards
PHYSICAL CHEMISTRY Peter Atkins
PILGRIMAGE Ian Reader
PLAGUE Paul Slack
PLANETS David A. Rothery
PLANTS Timothy Walker
PLATE TECTONICS Peter Molnar
PLATO Julia Annas
POLITICAL PHILOSOPHY David Miller
POLITICS Kenneth Minogue
POSTCOLONIALISM Robert Young
POSTMODERNISM Christopher Butler
POSTSTRUCTURALISM Catherine Belsey
PREHISTORY Chris Gosden
PRESOCRATIC PHILOSOPHY Catherine Osborne
PRIVACY Raymond Wacks
PROBABILITY John Haigh
PROGRESSIVISM Walter Nugent
PROTESTANTISM Mark A. Noll
PSYCHIATRY Tom Burns
PSYCHOANALYSIS Daniel Pick
PSYCHOLOGY Gillian Butler and Freda McManus
PSYCHOTHERAPY Tom Burns and Eva Burns-Lundgren

PURITANISM Francis J. Bremer
THE QUAKERS Pink Dandelion
QUANTUM THEORY John Polkinghorne
RACISM Ali Rattansi
RADIOACTIVITY Claudio Tuniz
RASTAFARI Ennis B. Edmonds
THE REAGAN REVOLUTION Gil Troy
REALITY Jan Westerhoff
THE REFORMATION Peter Marshall
RELATIVITY Russell Stannard
RELIGION IN AMERICA Timothy Beal
THE RENAISSANCE Jerry Brotton
RENAISSANCE ART Geraldine A. Johnson
REVOLUTIONS Jack A. Goldstone
RHETORIC Richard Toye
RISK Baruch Fischhoff and John Kadvany
RITUAL Barry Stephenson
RIVERS Nick Middleton
ROBOTICS Alan Winfield
ROMAN BRITAIN Peter Salway
THE ROMAN EMPIRE Christopher Kelly
THE ROMAN REPUBLIC David M. Gwynn
ROMANTICISM Michael Ferber
ROUSSEAU Robert Wokler
RUSSELL A. C. Grayling
RUSSIAN HISTORY Geoffrey Hosking
RUSSIAN LITERATURE Catriona Kelly
THE RUSSIAN REVOLUTION S. A. Smith
SCHIZOPHRENIA Chris Frith and Eve Johnstone
SCHOPENHAUER Christopher Janaway
SCIENCE AND RELIGION Thomas Dixon
SCIENCE FICTION David Seed
THE SCIENTIFIC REVOLUTION Lawrence M. Principe
SCOTLAND Rab Houston
SEXUALITY Véronique Mottier
SHAKESPEARE'S COMEDIES Bart van Es
SIKHISM Eleanor Nesbitt
THE SILK ROAD James A. Millward
SLANG Jonathon Green
SLEEP Steven W. Lockley and Russell G. Foster
SOCIAL AND CULTURAL ANTHROPOLOGY John Monaghan and Peter Just
SOCIAL PSYCHOLOGY Richard J. Crisp
SOCIAL WORK Sally Holland and Jonathan Scourfield
SOCIALISM Michael Newman
SOCIOLINGUISTICS John Edwards
SOCIOLOGY Steve Bruce
SOCRATES C. C. W. Taylor
SOUND Mike Goldsmith
THE SOVIET UNION Stephen Lovell
THE SPANISH CIVIL WAR Helen Graham
SPANISH LITERATURE Jo Labanyi
SPINOZA Roger Scruton
SPIRITUALITY Philip Sheldrake
SPORT Mike Cronin
STARS Andrew King
STATISTICS David J. Hand
STEM CELLS Jonathan Slack
STRUCTURAL ENGINEERING David Blockley
STUART BRITAIN John Morrill
SUPERCONDUCTIVITY Stephen Blundell
SYMMETRY Ian Stewart
TAXATION Stephen Smith
TEETH Peter S. Ungar
TERRORISM Charles Townshend
THEATRE Marvin Carlson
THEOLOGY David F. Ford
THOMAS AQUINAS Fergus Kerr
THOUGHT Tim Bayne
TIBETAN BUDDHISM Matthew T. Kapstein
TOCQUEVILLE Harvey C. Mansfield
TRAGEDY Adrian Poole
THE TROJAN WAR Eric H. Cline
TRUST Katherine Hawley
THE TUDORS John Guy
TWENTIETH-CENTURY BRITAIN Kenneth O. Morgan
THE UNITED NATIONS Jussi M. Hanhimäki
THE U.S. CONGRESS Donald A. Ritchie
THE U.S. SUPREME COURT Linda Greenhouse
UTOPIANISM Lyman Tower Sargent
THE VIKINGS Julian Richards
VIRUSES Dorothy H. Crawford
WATER John Finney
THE WELFARE STATE David Garland
WILLIAM SHAKESPEARE Stanley Wells
WITCHCRAFT Malcolm Gaskill
WITTGENSTEIN A. C. Grayling
WORK Stephen Fineman
WORLD MUSIC Philip Bohlman
THE WORLD TRADE ORGANIZATION Amrita Narlikar
WORLD WAR II Gerhard L. Weinberg
WRITING AND SCRIPT Andrew Robinson

Alexander C. Diener and Joshua Hagen

BORDERS

A Very Short Introduction

OXFORD
UNIVERSITY PRESS

Oxford University Press is a department of the University of Oxford.
It furthers the University's objective of excellence in research,
scholarship, and education by publishing worldwide.

Oxford New York
Auckland Cape Town Dar es Salaam Hong Kong Karachi
Kuala Lumpur Madrid Melbourne Mexico City Nairobi
New Delhi Shanghai Taipei Toronto

With offices in
Argentina Austria Brazil Chile Czech Republic France Greece
Guatemala Hungary Italy Japan Poland Portugal Singapore
South Korea Switzerland Thailand Turkey Ukraine Vietnam

Published in the United States of America by
Oxford University Press
198 Madison Avenue, New York, NY 10016

Library of Congress Cataloging-in-Publication Data
Diener, Alexander C.
Borders: a very short introduction / Alexander C. Diener and Joshua Hagen.
p. cm.
Includes index.
ISBN 978-0-19-973150-3 (pbk. : alk. paper)
1. Boundaries. 2. Borderlands. 3. Boundary disputes.
4. Human territoriality—Political aspects. 5. Human geography.
6. Political anthropology. 7. International relations.
I. Hagen, Joshua, 1974– II. Title.
JC323.D54 2012
320.1'2—dc23 2011052918

7 9 8 6

Printed in Great Britain
by Ashford Colour Press Ltd., Gosport, Hants.
on acid-free paper

Contents

List of illustrations and maps

Acknowledgments

We would like to thank everyone at Oxford University Press, especially Nancy Toff and Sonia Tycko, for their efforts in bringing this book to fruition. We also appreciate the helpful comments provided by the anonymous reviewers. Much of Alexander Diener's contribution to this manuscript was written while he was the Senior Scholar in Eurasian Studies at the Institute for European, Russian, and Eurasian Studies at George Washington University's Elliott School of International Affairs. Joshua Hagen received support from Marshall University and the Alexander von Humboldt Foundation during the writing of this book. Finally, we thank our families, Joy Ann Souligny and Rachel, Sabina, and Oliver Hagen, for their continuing support.

Chapter 1
A very bordered world

We live in a very bordered world. The daily news is filled with controversies concerning the political, cultural, and economic borders that crisscross the Earth's surface. Borders are central features in current international disputes relating to security, migration, trade, and natural resources. They also factor prominently into local debates over land use and property rights. Regardless of the scale, it is clear that humans draw lines that divide the world into specific places, territories, and categories. We are "geographic beings" for whom the creation of places, and by consequence the process of bordering, seems natural. But borders are not "natural" phenomena; they exist in the world only to the extent that humans regard them as meaningful. This Very Short Introduction seeks to present borders or geographic boundaries in a manner that reveals their socially constructed quality and by consequence our capacity to use them, change them, or even abolish them. We explore the theoretical and empirical complexity of borders and bordering processes while revealing the broad effects these phenomena have on our daily lives. Truth be told, most people cross hundreds of geographic boundaries on a daily basis. Some are formal borders demarcating ownership or the limits of governmental authority, while others are symbolic or informal associations of places with social groups or ideas.

Our daily routines provide a simple example. Typical mornings involve spaces specifically designed to limit access, such as bedrooms and bathrooms, as well as more open spaces like kitchens and dining areas. A trip to work usually requires leaving private property and passing through various public spaces, neighborhoods, or municipalities. Workplaces are also divided into spaces designated for specific purposes (offices, lunchrooms, factory floors, etc). The borders that define these various spaces, whether familial, social, economic, or political, address issues of access, mobility, and belonging in different ways. For example, a factory gate is intended to restrict access to certain people, whereas the entrance of a retail store is designed to lure people inside. This highlights the seemingly contradictory role of borders as bridges, gateways, and meeting points or barriers, obstacles, and points of separation.

In addition to influencing movement, geographic borders also define spaces of differing laws and social norms. In this way, borders create and signify varied legal obligations, social categories, and behavioral expectations for different areas. Returning to our example, some work spaces might require a hardhat or hearing protection, while lowered voices and business attire are expected in other areas. Signs restricting access to employees or paying customers signal authority over space and differentiate between groups of people. These mundane examples reveal the diverse role of borders as dividers of space, symbolic markers of control, and social processes of daily life. Reflecting their significance, borders have become a focus of study across the social sciences and humanities.

This book demonstrates the importance of borders as a topic of study by reviewing how the bounding of space has been an essential component of human activity for millennia. Beginning as fuzzy zones between tribal groups, the phenomenon of "frontiers" eventually evolved to encompass the transitional spaces between walled cites or imperial realms. Following

Europe's wars of religion during the sixteenth century culminating in the Peace of Westphalia in 1648, frontiers across Europe were gradually morphed into seemingly rigid lines dividing nation-states. This model of organizing political space was subsequently exported to the rest of the world through European colonial conquest, most prominently during the eighteenth and nineteenth centuries. Efforts at bureaucratic control within these states, and at times ethnic, cultural, or religious differences, compelled the further division of space into provinces, counties, townships, cities, reservations, and other administrative districts. In addition to these formal governmental structures, countless informal social boundaries, such as gendered spaces, gang territories, gated communities, and ethnic neighborhoods, are also expressed spatially. All of these have some type of formal or informal border marking them apart from other political and social spaces.

1. Workers erect a new fence along the U.S.-Mexico border near El Paso, Texas, ca. 2011.

Ultimately, the world has become crisscrossed with such a variety of geographic boundaries that they often appear natural and timeless. Yet reality is more complicated. Although the bounding of space may be common in human social organization, borders are *not* themselves strictly natural phenomena. Or put another way, humans may be geographic beings predisposed to spatial organization, but how we structure territory, and to what end, has evolved quite radically over time reflecting changing political, social, and economic contexts. The theoretical foundations of bounded space encompass a wide range of scholarly perspectives. Although the entirety of this vast and growing field cannot possibly be covered here, we hope this broad survey of border history and contemporary border research inspires greater awareness and further study of these topics among scholars, students, and general readers.

Territory, sovereignty, and borders

The primary function of geographic borders is to create and differentiate places. In other words, borders separate the social, political, economic, or cultural meanings of one geographic space from another. While the world is replete with various geographic boundaries, the institutional phenomenon of borders is most commonly associated with the idea of territory. In many languages, such as French, Spanish, and Italian, the term territory is synonymous with "place" or "space." However, in the English language, social science usage of "territory" generally references the jurisdiction of countries (or states). This book defines territory as a geographic area intended to regulate the movement of people and engender certain norms of behavior. The process of creating territories requires some mode of territoriality.

Territoriality is the means by which humans create, communicate, and control geographical spaces, either individually or through some social or political entity. Modes of territoriality have varied significantly over time and across space resulting in diverse practices of bordering. These range from the placement of permanent markers to the performance of intermittent ceremonies and from the precise demarcation of sharp lines to the broad definition of transitional zones. Therefore, territoriality and practices of bordering are neither constant nor consistent but rather highly contingent and adaptable. Though pervasive throughout recorded history, the root causes of territoriality have been long debated by scholars. Some have favored sociobiology or primordialist approaches believing that territoriality stems from an *a priori* instinct. In this view, social groups instinctively seek territorial control to secure resources necessary for survival. This suggests humans are subject to perpetual "survival of the fittest" contests as groups seek to control territory, secure resources, and deny access to competing groups. Such a perspective is highly problematic. Although animals exhibit territoriality, for example, in marking out hunting ranges, attempts to link human territoriality to mere instinct unduly reduces a far more complex process to a natural reflex. Human place-making and territoriality differ from that of animals in two distinct ways.

First, territorial control is not, nor has it ever been, the sole means by which humans enact political power. Countless forms of de-territorialized "authority" (the legitimate exercise of power) have existed throughout history and continue to exist today. Contemporary examples include various religious and social movements, as well as nongovernmental organizations relating to environmentalism, human rights, and feminism that propagate their ideologies as universal and claim authority across space, class, and various forms of identity. The global influence of certain businesses, such as microprocessor giant Intel, could also be considered a form of de-territorialized authority since these technologies clearly transcend territorial boundaries.

The second manner in which human territoriality differs from that of animals relates to the evolution of human territoriality.

Unlike animals, human spatial thinking has manifested in very different ways over time. For example, frontiers, or zones of limited rule of law, were once commonplace in the world but today are conspicuous and rare. Also, some human communities did not develop a conception of land "ownership" until forced to do so by groups that had done so. While territoriality existed among both groups, it is clear that the phenomenon manifested very differently.

Consideration of the evolution of human spatial thinking has catalyzed alternative theories as to the causes of territoriality. Some scholars favor a constructivist approach that rejects the environmental determinist notions of primordialists. Constructivists contend that territoriality results from historical contexts, practical needs, and geopolitical contingencies. These scholars suggest that determinations of "us" and "them," "insiders" and "outsiders," and "in place" and "out of place" are not related to what we commonly identify as innate categories such as race and ethnicity, or even cultural characteristics such as language or religion, but are formed through unequal power relations within and between social systems. Territoriality thereby serves as a social mechanism for this control, driving the process of defining what is "ours" in opposition to what is "theirs." By demarcating and defending territory, groups control specific spaces and resources in an effort to regulate extraterritorial practices, such as entry and exit, and intraterritorial practices, such as social hierarchies and governance. Regardless of its origin, territoriality has become institutionalized in the modern era with the effect of naturalizing the overtly social processes of bordering.

As manifestations of territoriality, borders provide a means to assign things to particular spaces and regulate access into and/or out of specific areas. This innately social and political process links to the idea of ownership or rightful and permanent possession of land. Over recent centuries, the development of dramatic disparities in power within and between human groups has given rise to the concepts of sovereignty and jurisdiction. Though rather recent inventions, these concepts were central to defining the limits of a nation's power and establishing borders as the organizing principles for the modern state system.

Sovereignty can be defined as the exercise of supreme authority and control over a distinct territory and its corresponding population and resources. Jurisdiction refers to a bounded area within which the authority of a particular person, group, or institution is legally recognized. Though similar in form and function, jurisdiction generally relates to a smaller area of authority within a higher order sovereign entity. Both are very complex ideas that help define the spatiality of governance (state, ethnic territory, province, municipality, etc.) and the nature of that control. In some cases, control may be benign or beneficial by offering sanctuary and security. In other cases, the control may be oppressive or violent constituting a prison to those within or a fortress against those outside.

Sovereignty and jurisdiction signify at least some form of popularly recognized authority over a bounded territory while also disguising the often violent origins of such rule. They mask the processes by which other "spatialities," or modes of human interaction with space, and identities were plowed under. Because identity is not static and spatial patterns of interaction and exchange are fluid, complete territorial sovereignty and jurisdiction (an "air-tight seal") are never fully achievable. Not even the famed Iron Curtain completely blocked out external ideas and goods. As such, territorial sovereignty and jurisdiction remain, in conjunction with national territoriality, both causes and responses to much intra- and interstate tension and conflict. For example, Israeli settlers have moved beyond the margins of their sovereign state territory in an effort to claim land they believe to be theirs and enhance security for the country's core. These acts serve to mobilize Palestinian resistance and countermeasures to express sovereignty and enhance their own security. In this example, and many others, borders are active forces helping constitute dynamics of order and disorder across varied global and local landscapes.

Today, perhaps more than ever, the rise of cross-border processes, patterns, and problems, often lumped together under the term "globalization," challenge established notions of territorial sovereignty. For example, environmental impacts resulting from pollution or climate change transcend bounded spaces. One country's efforts to "go green" may be undermined by its dirtier neighbors' industrial practices that invariably affect common air and water quality. In economics and business, transnational corporations increasingly benefit from common markets (e.g., the European Union [EU] and North American Free Trade Agreement [NAFTA]) and lower tariffs as they build commodity and supply chains that span the world. Global networks of information, innovation, and education now link the East to the West and the North to the South so intimately that the gap between "haves" and "have-nots" is a conspicuous fact of life. Numerous supranational institutions and private groups work across borders to narrow this gap, but they are not always successful. Dual, multiple, or changing citizenship is also increasingly sought among the highly mobile members of various societies. This includes both wealthy cosmopolites jet-setting between global cities, and migrants fleeing war, famine, environmental degradation, oppression, or poverty. In addition to the various migrants crossing borders in pursuit of opportunity, a range of nonstate combatants, like terrorists, pirates, and mercenaries, cross borders to engage in illicit and destructive activities.

All this transborder activity has inspired states throughout the world to enhance their border security in an effort to better manage the flows into and out of their territories. Examples include the United States constructing hundreds of miles of fencing along the Mexican border; India fencing its 2,500-mile (4,000-km) border with Bangladesh and 1,800-mile (2,900-km) border with Pakistan; Pakistan building fences and laying minefields along sections of the Afghanistan border; and Iran walling its 430-mile (700-km) border with Pakistan. Israel is constructing a 470-mile (760-km) security barrier around many Palestinian areas in the West Bank, while shorter fences are being built along Israel's borders with Gaza and Egypt. Beyond these highly publicized examples, numerous other countries, including China, Greece, Kuwait, Morocco, Saudi Arabia, Spain, Thailand, Uzbekistan, and the United Arab Emirates, have launched new fence construction projects. In sum, as of 2011, approximately 12,500 miles (ca. 20,000 kilometers) of the world's borders are marked with walls or fences and an additional 11,000 miles (ca. 18,000 kilometers) host noticeable security enhancements such as surveillance technology and patrols.

These contradictions suggest that borders may be conceived as spatial practices comprised and maintained by a continual negotiation between the boundedness of territories and cross-border flows of people, goods, capital, and information.

2. Israel has built massive walls and other barriers to seal off the Palestinian West Bank.

Scholars from the past and present have explored the utility, consequences, and contradictions inherent within the concept of borders and bounded territories. Writers from Plato (*Laws*, 360 BCE) and Aristotle (*Politics*, 350 BCE) to Thomas Hobbes (*Leviathan*, 1651 CE) and Adam Smith (*The Wealth of Nations*, 1776 CE) to Thomas Friedman (*The World Is Flat*, 2005 CE) and Harm de Blij (*The Power of Place*, 2008 CE) provide testimony that geopolitical systems are neither completely closed (the ideal of rigid territorial sovereignty) nor open (an "end of geography" brought on by flows of globalization).

This tension between openness and closedness has been articulated numerous times since philosophers first debated the ideal society. In Plato's *Laws*, Sparta is presented as a model polity, which emphasizes "security" in pursuit of the ultimate goals of virtue and happiness. "Opportunity" in the form of maritime expeditions, distant resource exploration, and profit seeking through territorial expansion was considered best constrained. In contrast, Aristotle's *Politics* critiques Plato by calling for balance between security and opportunity. From this perspective, isolation is an unattractive option since the maintenance and perpetuation of the polity require engagement with external sociopolitical entities.

Over time, oscillation between the ideals of security and opportunity helped form the territorial state model that characterizes the contemporary world. By the time the nation-state system was largely codified in Europe through the Peace of Westphalia, these same European states had developed a range of technological advantages that would facilitate conquest of much of the world's land area. These states pursued strategies focusing on "opportunity" in the form of colonial territorial expansion only to gradually reemphasize "security" in the form of rigid sovereignty over the lands they conquered. Reflecting territoriality's tendency to proliferate interactively, this ultimately led to territorial ambitions among indigenous groups, which catalyzed both decolonization and nationalization throughout the world. As a result of these processes, the world's land area was gradually carved into the sovereign nation-states we have today. Although borders were often conceptualized as simple barriers and dividers of space, these were never and are certainly not today the only roles that borders play.

Global headlines offer poignant evidence for the contradictory nature of contemporary border regimes. While pockets of isolated sovereignty (enclaves and exclaves) within or beyond the borders of Spain, Morocco, Angola, Armenia, Azerbaijan, Uzbekistan, and Kyrgyzstan have, in respective incidents, nearly sparked war in recent years, the EU's Schengen Regime, a program for visa-free travel among member countries, has diminished travel restrictions among many states and brought them into closer confederation. Where concerns over terrorism, drug trafficking, and illegal migration have heightened border controls throughout the world, multinational companies, tourists, and skilled migrants simultaneously enjoy greater border permeability. Indeed, passports and visas offer sophisticated means of monitoring human mobility. This emphasis on surveillance constitutes an effort to balance the opportunity offered by increased international exchange with desires for greater security. It also highlights the need for cross-border cooperation in the management of issues as broad ranging as tourism, labor migration, environmental conservation, and crime. Since borders possess two distinct sides, the pursuit of border security and effective management of trans-state practices require coordinated bi- and/or multilateral efforts. It is for this reason that moral and ethical tensions resonate at the junction of state sovereignty and globalization. This resonance challenges rigid perspectives of borders as either immutable barriers or fading anachronisms. In fact, new spaces of sovereignty and authority are emerging and shattering the fictive, nested hierarchy of territorial jurisdictions, starting with private homes and ending at the nation-state.

While external state borders are central to a variety of issues, new economic, social, and political realities are producing new forms of bordering and alternative spatial realities manifest at the substate level. Voting districts, census tracts, municipal boundaries, and any number of other bureaucratic divisions of space, along with unofficial boundaries of socioeconomic and cultural differences, increasingly constitute tangible landscapes of authority and power. These civil hierarchies and the varied spatialities they foster play central roles in shaping individual and group identity. Our relative position within these hierarchies (this side of the tracks or that, this neighborhood or that, this city or that, this country or that) shapes our sense of self, belonging, and life trajectory. The aspirations, opportunities, and realities of people are conditioned, in large part, by the varied degrees of power vested within each geographically defined unit.

Nevertheless, it should be noted that while most people would feel relatively comfortable revising a census tract or local park boundary, the lines partitioning the colorful collage of countries on world maps convey an air of sanctity. Different perceptions of the significance and permanence of geographic boundaries are not accidental. International borders have been purposely constructed and represented to appear as though they derive from some higher logic. They are, however, no more natural or logical than obviously contrived school zones or electoral districts.

Whether based on seemingly objective criteria, such as rivers or lines of latitude, or appearing convoluted and artificial, all borders are delineated in accordance with human biases, beliefs, and assumptions. Territorial disputes and competing border practices are often so intractable because individuals or groups are supremely confident in the justness of their respective claims. As such, every geographic boundary is a symbolic representation and practical embodiment of human territoriality. This has been

economics, and sociology. Nevertheless, beginning slowly in the 1960-70s with decolonization, gaining momentum through the 1980s with the lessening of Soviet global power, and booming in the 1990s with the (re)emergence of new states from the collapsed USSR, the topic of borders edged back into academic writing as scholars questioned the overt emphasis on "security" rather than "opportunity." Border studies gained even greater popularity with the advance of neoliberal economic thought, the consolidation of the European Union, and the resurgence of protectionist ideologies in reaction to the spate of terrorist attacks in the early 2000s.

The contradictory and controversial nature of contemporary borders has provided a fertile ground for research as evidenced by the establishment of an array of research centers and a growing international community of border scholars. Examples of contested boundaries and border-related issues abound but ironically exist today amid a growing sense that borders are diminishing in importance. Greater mobility among general populations within and between various countries and provinces, the development of regional and supranational institutions, and increased cross-border economic activity point to at least partial de-territorialization of human identity, state sovereignty, and national economies. Some even declare the end of geography, meaning that a world of flows is erasing the world of places and will ultimately give rise to a homogenized, borderless landscape.

Yet most scholars see something quite different on the horizon. Far from the death knell for states as the primary organizational unit of global political space, one cadre of academics and analysts projects a scaling up to larger geopolitical structures that would still involve borders lining their edges. Others foresee a scaling down of territorial organization into overlapping and less rigidly defined local identities and concepts of sovereignty. Both alternatives entail new functions of territoriality and geographic boundaries rather than their abolition.

Contrary to such bold predictions, the effects of globalization are so wide-ranging that a singular effect on human spatiality is unlikely. Indeed, history reveals alternating patterns of border upheaval and stability, as well as permeability and rigidity. For example, the recurring conflicts among European powers from the Seven Years War to the Napoleonic wars to the World Wars marked a period of fluidity and instability as the contemporary political map coalesced. During the Cold War, however, international borders remained divisive issues but appeared static and fixed against the backdrop of superpower confrontation.

The collapse of this bipolar international system along with its figurative and literal border landmarks—the Iron Curtain and the Berlin Wall—resulted in renewed border instability across the former communist sphere. With its demise, the Soviet Union bequeathed fifteen new states to the world map. Yugoslavia and Czechoslovakia also fractured into new sovereign states and, whether desirable or not, this process may not be complete. Albanian-speakers in Kosovo have proclaimed their nation-state status, while Russia has occupied portions of Georgia in support of independence for South Ossetians and Abkhazians. Interestingly, countries as diverse as China and Spain share a reluctance to recognize new claims to sovereignty for fear of fueling national self-determination movements among their own dissatisfied minority groups (Tibetan and Uighur, Basque and Catalan respectively). The fracturing of empires and cobbled-together federated states, however, covers just some of the recent border-related headlines. Ironically, other headlines involve the consolidation of states into larger supranational organizations, such as the European Union or calls for some type of pan-Islamic caliphate. Yet even in these divergent examples, borders would not disappear but continue to surround these smaller or larger territorial units and enact territoriality in no lesser degree than the nation-state entities from which they are made.

This discussion makes clear that there is no single trend toward either consolidation or fragmentation. Some parts of the world are becoming more fragmented, while others are experiencing greater integration. Borders remain significant amid the fluidity of globalization because it is the existence of borders that put the "trans" (i.e., to cross, breach, or span) in processes of transnationalism and transmigration. Though potentially changing in nature with neoliberal capital flows and cyber community linkages across space, territoriality remains ubiquitous and unlikely to fade any time soon.

We are, nevertheless, experiencing a transition of human spatiality and borders. Contradictory border processes generate great anxiety among those tasked with adapting to this evolving spatial reality. Policymakers struggle to strike a precarious balance between facilitating cross-border trade and investment, and managing the post-9/11 security environment. It seems that for every initiative seeking to "soften" borders, facilitating a freer flow of people and materials, there are an equal number intended to "harden" borders, requiring new forms of documentation and constructing new physical barriers. These movements toward the physical materialization of borders must also contend with the new realities of cyberspace that facilitate novel human connections by reducing the friction of distance while simultaneously providing a forum for confrontation, discrimination, and hatred. The human capacity to employ borders to filter flows into and out of territory is central to this new era of shifting spatiality.

A very short introduction to borders

Borders are integral components of human activity and organization. As such, we are compelled to deepen our understanding of their role as areas of opportunity and insecurity, zones of contact and conflict, sites of cooperation and competition, places of ambivalent identities and aggressive assertions of difference. These dichotomies may alternate with time and place, but more interestingly, they often coexist. We must come to terms with the means by which borders structure our lives, while simultaneously lessening their perpetuation of antagonistic difference and apathetic indifference. We must find a way to harness their ability to catalyze belonging and identity but diminish their propensity for exclusion and the creation of "others." We must confront the increasingly transportable and multiscalar nature of territory and engage the multilayered role of borders in a variety of social settings. We must tackle ethical questions—For whom are borders constructed? By whom? And to what ends?

Borders require further study both from the top down and from the bottom up, from the state scale and from the local scale, as they are perhaps the most obvious political geographic entities in our lives. Ultimately, the lived experience of borders reminds us that their opacity is as important as their transparency. We must be mindful of borders' capacity to evolve in role and change in nature. Sites of cooperation can become sites of contestation, and vice versa. This book in no way represents a comprehensive treatment of borders. In fact, it is challenging to write a Very Short Introduction on such a diverse and complex topic. Yet it is for these reasons that borders are and will remain such important factors shaping our world.

Chapter 2
Borders and territory in the ancient world

Modern political maps confer a sense of permanence to the world's states and their borders. Indeed, a major objective of the modern social sciences has been to construct historical lineages that project contemporary political entities, such as France, China, and Iran, and their sovereign territories as originating in prehistory. Yet the modern political map and its underlying territorial assumptions are relatively recent developments. This is not to suggest that premodern societies and polities lacked notions of territoriality or borders; rather, they tended to conceive of these things and arrange social and political space in ways that differ from contemporary ideals and expectations.

Territoriality among hunter-gatherers

The earliest humans lived within small bands of nomadic foragers, known as hunter-gatherers. Migrations of these groups were generally patterned by changing seasons and environmental conditions that affected the availability of food and other resources. Given their nomadic way of life, one might assume these early peoples lacked notions of borders, territoriality, or land ownership. Yet research on contemporary hunter-gatherer groups strongly suggests that these prehistoric humans possessed surprisingly elaborate territorial arrangements. Rather than wandering aimlessly, hunter-gatherer groups likely operated within relatively stable local or regional foraging ranges, or what might be best described as networks of foraging sites. These networks were shaped by extended kinship or alliance relationships, religious beliefs, and ecological conditions. Yet notions of ownership and norms regarding access to territory and resources varied considerably.

Some groups invested considerable energy in claiming and maintaining exclusive access to specific hunting and foraging areas. For example, the Tsimshian, Coast Salish, and many other groups in North America's Pacific Northwest developed intricate social systems focusing on ownership of and access to resource sites. Although punishment for trespassers could be severe, property-owning lineages were expected to share their bounty through ritual celebrations. The Veddah peoples of Sri Lanka also marked and rigorously safeguarded specific territories. Each band was expected to be self-sufficient within its own hunting area, and trespassing was strictly prohibited in nearly all circumstances. In areas lacking an obvious natural feature, the Veddahs would carve symbols into tree trunks to mark the border. Although these and other hunter-gatherer groups may have shared a notion of exclusive territoriality, the concept of ownership resided with individual households in some groups, while others regarded broader familial lineages as owning the land in common.

Other hunter-gatherer cultures adopted alternative territorial strategies that controlled resources by emphasizing social cohesion and reciprocity. The !Kung bands of southern Africa had rather vague borders for their foraging ranges, often coinciding with natural landmarks, and made little effort to deny access to other bands. Indeed, resources were not considered to be owned until gathered, and goods were to be shared equally once collected. Bands had considerable freedom to access resources in neighboring ranges, especially with those bands sharing kinship ties. Such visits were frequent and generally welcome, so long as the visiting group first sought permission and shared some portion of the collected resources with the host band. Similar hospitality was expected in return. A comparable system was common among many Aboriginal groups in Australia where fluid membership and kinship ties also encouraged territorial and resource sharing among adjacent bands. Each band may have had a specific foraging network, but these areas were not considered exclusive and often overlapped with those of neighboring groups. In other cases, Aboriginal ranges were bordered by buffer areas, which were used infrequently and apparently not claimed by any group.

A great deal of variation and combination existed between the two main strategies discussed earlier as foraging bands merged or divided in response to population changes, resource availability, or personal disagreements. Although it is tempting to think of ancient hunter-gatherer groups as nonterritorial since clear borders seem absent among their modern-day counterparts, these early groups most likely featured complex strategies for regulating membership, territory, and resources that intermingled notions of exclusivity and reciprocity contingent upon animistic religious practices, kinship customs, and resource availability.

State formation in antiquity

Starting around the tenth millennium BCE, hunter-gatherer bands gradually transitioned to settled agricultural communities. This so-called Neolithic Revolution was triggered by the domestication of plants and animals that allowed the development of more reliable and bountiful food production. This first occurred in an area of productive farmland stretching across much of modern-day Iraq and Syria known as the Fertile Crescent. Settled agricultural communities apparently arose independently in parts of India, China, Africa, and the Americas by the third millennium BCE. In addition to permanent settlements, this shift from hunting-gathering to farming had far-reaching consequences for human history, including the development of writing, irrigation, architecture, government, and socioeconomic specialization.

The shift also provided a precondition for the emergence of new political organizations of territory. Modest settlements and later city-states ruled by local chiefs or kings gradually developed, especially along fertile river valleys.

The Sumerians were the earliest group to transition to a city-state culture for which we have relatively good information. By around 4000 BCE, the Sumerians began to coalesce into a few dozen significant city-states scattered along the Tigris-Euphrates river basin. A city-state is a sovereign polity encompassing a relatively small area consisting of an urban core and surrounding farmland. In the case of Sumer, each city-state was independent and ruled by its own king, who represented the city's patron deity. These city-states relied on the agricultural surpluses produced in the surrounding fields to sustain the city's population. If a king wished to expand his city, which might increase his tax revenues, he had to acquire more farmland. In ancient Sumer, this generally meant investing in expanding irrigation canals to farm new lands or simply annexing established fields from neighboring city-states.

Given the importance of adequate farmland for their survival, it is not surprising that Sumerian city-states were seemingly in near constant conflict. Some of the Sumerian kings were quite successful. Yet these conquests seemed more likely to result in tributary payments from the defeated city-state rather than outright territorial annexation, and even these tributary arrangements appeared short-lived. Evidence for this comes from *stelae* (plural of *stele*), or commemorative stone slabs, which ancient rulers erected to record their achievements. While some were meant to commemorate important events, like a military victory or a ruler's accession to the throne, stelae were also commonly used to establish territorial claims and boundaries. Some marked internal administrative boundaries, while others represented external borders between neighboring kingdoms.

In ancient Egypt, for example, the pharaoh Akhenaten marked the limits of his new capital city at Amarna with a series of stelae. In addition to indicating the city's limits, the stelae were also inscribed with religious dedications, which divided the sacred space of the capital from the rest of the kingdom. Other pharaohs used stelae to mark the outer limits of their realm, as did Senusret III, who erected stelae to designate Egypt's southern border with Nubia. Yet the stelae were not intended to mark a clearly defined linear border. Instead they were located inside frontier fortresses and made general claims over the surrounding territories rather than marking an absolute divide in sovereign territory. A single linear border in the modern sense did not exist. Rather the inscriptions on these stelae and nearby statues suggest these fortresses served more as "checkpoints" where government officials could regulate the movement of people and goods along the Nile, a vital travel corridor. Stelae served similar functions for the Olmec and Mayan civilizations in Central America. These stelae have proven to be important sources for understanding these ancient cultures and their conceptions of territoriality and borders.

The earliest known border conflict is recorded on the so-called Stele of the Vultures that describes a dispute over farmland between the prominent Sumerian city-states of Lagash and Umma during the twenty-fifth century BCE. The Stele of the Vultures and other ancient inscriptions provide a general history of the dispute. The top Sumerian deity Enlil had originally set the border between Lagash and Umma, but the two cities disputed its actual location. Acting as some type of outsider arbitrator or even divine intermediary, Mesalim, the king of the city-state Kish, later surveyed the border and erected a stele inscribed with his decision. Despite this, subsequent rulers in Lagash and Umma quarreled incessantly over the fields and water rights in the area. Eventually, the ruler of Umma removed the stele and seized all the land. After a number of years, Eannatum, ruler of Lagash, retaliated and defeated Umma. Eannatum restored the original boundary stele and dug an irrigation canal to mark the border. He also erected a second stele outlining a harsh peace treaty imposed upon Umma. This treaty specified which fields belonged to Lagash and Umma, obliged Umma to maintain a buffer zone on its side of the border, and required payments of tribute to Lagash. Finally, Eannatum sought to affirm the border's sacred sanction by building shrines to Enlil and other deities. These stelae and shrines marked only a portion of the border, but they contained inscriptions describing the entire course of the border.

The city-state would remain a prominent form of territorial and political organization for centuries, but broader trends were already emerging that enabled the formation of larger and more centralized structures. As agricultural productivity improved, city-states were able to grow demographically, specialize economically, and support expanded government bureaucracies and standing militaries. These changes made it possible for ambitious city-state rulers to conquer and annex neighboring city-states, rather than simply demanding payments of tribute.

Again, the ancient Fertile Crescent provides the earliest known example in Sargon the Great, ruler of Kish. Although ruling dynasties in antiquity were continually being established and overthrown, Sargon managed to create and sustain an expansive and centralized system of direct rule. Following a series of military victories, Sargon established the Akkadian Empire, which controlled most of the formerly independent city-states of Mesopotamia during much of the twenty-fourth and twenty-third centuries BCE. It is often recognized as the world's first empire.

An empire is a sovereign political entity, usually governed by a hereditary monarch, and spanning several different regions and peoples, which had not traditionally been under a single ruler. Empires require more complex and extensive administrative structures to govern effectively because of their greater size and diversity than traditional city-states. This often entailed the delegation of authority to lower-level governors and administrators who act on behalf of the central authority. In the case of the Akkadian Empire, Sargon and his successors appointed governors, or *énsi*, to administer the areas beyond the capital city district. In many cases, the Akkadians left the original rulers in charge, but conquered city-states were now directly subordinate to the central imperial authority. In lieu of paying tribute but otherwise retaining considerable autonomy, the *énsi* organized the direct payment of taxes to the royal treasury and were obligated to enforce the king's will.

Despite their apparent sophistication, the Akkadian Empire and its successors in ancient Mesopotamia proved vulnerable. Internally, regional governors often rebelled to reassert their traditional city-state autonomy. The *énsi* of Sumer were notorious for rebellion against their Akkadian masters. Externally, the threat of attack may have spurred neighboring city-states toward greater cooperation and even centralization of political power to defend against Akkadian or Sumerian attacks. This may have happened among the Elamites in what is now southwestern Iran. Prosperity within empires likely made them more tempting targets for neighboring city-states and empires as well as nomadic groups. Indeed, an invasion by the Gutians, nomads from the northern parts of modern-day Iran, eventually marked the end of the Akkadian dynasty.

This brief foray through ancient Mesopotamia highlights the three main modes of political-territorial organization that dominated the ancient world: migratory tribal groups or chiefdoms (i.e., Gutians), city-states (i.e., Lagash), and empires (i.e., Akkadian). A fourth possibility was some type of federation encompassing various tribes or city-states banding together to face a common opponent, although these types of structures tended to be relatively short-lived. Over the following centuries, successive empires were established, expanded, and collapsed, including the Assyrian, Persian, and Roman empires. City-state cultures thrived among the Greeks, Phoenicians, and Philistines. Nomadic and seminomadic groups, like the Hyksos, Aramaeans, and Amorites, continually plundered and occasionally conquered these city-states and empires. It is important to resist reading this as a simple progression, or "scaling up," from simple marauding bands to modest city-states to larger complex empires. Rather, the ancient world's political scene fluctuated back and forth between these varied territorial structures in response to changing environmental conditions, agricultural productivity, trade routes, military technologies, and the relative strength of neighboring groups. Despite changing contexts, each of these political entities had to develop some sort of strategy or system for the control and use of territory, people, and resources.

Nomadic groups

Ancient nomadic groups may appear similar to prehistoric hunter-gatherers, but such superficial similarities are misleading. These pastoral nomads developed systems of livestock-rearing adapted for areas with limited farming potential, such as semiarid regions of the Central Asian steppe or Africa's Sahel. Although little is known about the territorial organization among these groups in antiquity, such as the Scythians or Huns, we can infer much from such groups as the Kazakhs or Mongols, who retained nomadic lifestyles into recent times. Although they generally did not conceive of individual land ownership or that land had any intrinsic value, their reliance on animal husbandry shaped their political and territorial organization. Among these pastoral groups, territorial access was framed around the movement of livestock between seasonal grazing ranges, which were generally conceived as belonging to specific lineages or kinship groups.

The harsh Eurasian steppes had been home to various nomadic pastoralists for centuries. These rather decentralized groups could come together to launch plundering raids or repel outsiders encroaching on commonly held grazing areas. As a result, these nomads posed a continuing threat to the sedentary cultures stretching through China, the Middle East, and into Europe. They also constituted important trading partners and at times allies in campaigns against other nomadic groups or sedentary neighbors. Occasionally, powerful nomadic chieftains were able to assemble confederacies of tribes strong enough to conquer their nomadic and sedentary neighbors. The empire established by Mongol leader Chinggis Khan (Genghis Khan) was the largest of these and eventually covered much of the Eurasian landmass by the late thirteenth century CE. Chinggis Khan and his successors developed a more centralized administrative structure to govern this immense territory. While traditional land-use practices remained under the supervision of local tribal leaders, the khans appointed governors to oversee areas beyond the Mongol homeland. Instead of well-defined external borders, the Mongol Empire linked these far-flung administrative centers through an extensive network of roads, relay stations, and outposts, known as the *Yam*. The Yam served as arteries of information, trade, and military deployments and linked East, Southwest, Central, and South Asia as never before or since.

Although lacking rigid linear borders, nomadic peoples possessed notions of territoriality and made use of boundaries. Walled or fenced burial sites exist in a variety of nomadic cultural settings. These groups, however, tended to apply more fluid notions of territoriality at larger scales. For most mobile pastoralists of Eurasia, the idea of borders related more to ecological zones or simply distance. Archeological evidence of tool and other material exchanges (furs/hides, crafts, fabrics) suggests that the Bronze and Iron Ages saw a great deal of cultural mixing between forager and pastoralist cultures at the interface of forest, wetland, desert, mountain, and steppe. Long before the "Silk Road" traversed the plains of Central Asia, a "fur route" linked the steppe nomads to distant lands and settled peoples of the north. Although firm borders and direct claims of ownership appear less prominent among nomadic peoples, distinct markers of territorial presence and/or authority may be found in the burial sites, funerary monuments, and *balbals* (sometimes known as deer stones) that dot the Eurasian landmass. The geopolitical and social role that these monuments played continues to be hotly debated. The extent to which political organization among these groups required any territorial demarcation remains unclear. Pastoral economies have proven viable both within an imperial state structure and without any political hierarchy at all. From East Africa to Mongolia, one finds pastoralists shunning central political authorities at various points in history. While the claim that nomads lack a sense of territoriality is spurious, it is fair to say that their notions of territoriality were far more fluid than those of sedentary groups. It was, in many ways, the fluidity of nomadic territoriality that catalyzed specific architectural and sociopolitical adaptations among expanding sedentary populations.

City-states

City-states were modest in area and population, typically ranging from 100 to 1,000 square kilometers (259–2,590 square miles) and encompassing 5,000 to 25,000 inhabitants. In rare cases, they grew far larger, such as ancient Athens or pre-Columbian Tenochtitlan. Scholars have generally regarded city-states as either dead ends or transitional forms in the political evolution from simple tribal chiefdoms to complex empires and eventually modern states. These interpretations reflect the uncritical projection of modern-day territorial assumptions and practices back into prehistory and ignore the fact that city-states flourished as the dominant political organization in various parts of the world over long periods of time. In addition to their cultural and technological achievements, the Sumerian city-states managed to persist for nearly two millennia. Even when incorporated into the Akkadian Empire, the Sumerian cities revolted often and eventually reestablished their independence. This is testament to the city-state's enduring appeal and viability in the region. City-states also proved long-lived in other parts of the world. The city-state cultures of the Yoruba in West Africa, the Phoenicians throughout the Mediterranean, and the Nahuatl in the Valley of Mexico, for example, thrived for centuries.

Given their small size, these city-states may appear rather unsophisticated in terms of territorial organization and governance, but that was not the case. Rather than isolated, self-sufficient entities, ancient city-states tended to exist in spatial clusters possessing significant amounts of interaction and exchange. The city-state clusters functioned as hierarchical yet flexible networks of political, economic, and cultural cooperation and competition. These networks provided a major incentive for city-states to adopt modes of territorial organization, which would maximize opportunities while minimizing threats. In practice, this often meant acceptance of mutually beneficial material, technological, and cultural exchanges with neighboring city-states, while maintaining a vigorous defense against external threats to independence. Yet there were significant variations in the resources and importance invested in border creation and maintenance, as well as which types of borders assumed priority.

The city-states, or *poleis* (plural of *polis*), of ancient Greece are the best studied of the early city-state cultures. Much is known about the culture, politics, and economies of Greek city-states, but less effort has been made to understand how ancient Greeks organized territory. Indeed, scholarly attention seldom extends beyond the city's fortifications. This perspective is illustrated by the tendency to represent Greek city-states as mere dots without mapping out their territories. Yet there are numerous examples of disagreements over territory leading to war between city-states. It appears likely that most city-states, especially the larger ones, placed stone markers, known as *horos*, at locations where travel routes intersected their outermost territories. Religious sanctuaries were often located nearby as a way of providing divine justification for the city-state's territory. Travelers often marked border crossings by offering a sacrifice to the gods. Neutral zones commonly separated the borders of neighboring city-states.

By mutual agreement, this unclaimed territory was open for livestock grazing, but the establishment of fields or settlements was prohibited. The "border" wars among Greek city-states appear just as likely to center on disputes over the usage of this neutral territory rather than the actual location of the border. This neutral zone and the border's religious connotations help explain why relatively little territory was transferred between poleis as a result of warfare. Defeated city-states were more likely to be forced to ally with the victor or pay tribute than lose territory.

Unlike the Greek polis, the classical Mayan city-state, or *ahawlel*, generally lacked exterior walls, but considerable effort went into marking the borders between city-states. The process of border creation involved elaborate ceremonies after which participants would walk the boundary, marking it as they went. The process of dividing farmland within each city-state's territory was equally important. Each village had its own farming territories, which were then organized into parcels. The Maya did not buy or sell land, so these parcels were passed down through familial lineages, which maintained the borders of their farmlands through rituals that combined physically walking the border with ancestor worship.

Although some city-state cultures devoted significant time, labor, and resources to marking their territories, this was not universal. The Malay city-states, or *negeri*, of the fifteenth and sixteenth centuries, for example, seem to have done little to maintain borders. These Malay states were structured around river systems with the main urban center located at the river's mouth. The city's hinterland consisted of loosely controlled villages scattered upstream along the river and its tributaries. Because the territory away from the rivers was generally mountainous jungle and therefore poor farming land, it was sparsely populated and difficult to traverse. As a result, the main city at the river's mouth could dominate the flow of resources, goods, and people throughout the river basin without having to maintain extensive inland borders. It appears that the coastal centers did not even bother to implement direct rule over their hinterlands, preferring instead to leave local chiefs in charge. Instead of the compact and contiguous territories seen on modern maps, the territories of the Malay city-states are better imagined as networks of sites resembling a tree. The core coastal city-state was located at the base of the trunk while inland villages were scattered along the upper branches. Any trade between river basins would almost always mean passage through their respective urban cores.

Several factors explain these varied border strategies, including differences in environmental conditions, settlement and population densities, and agricultural and economic practices. For example, the areas beyond Greek city-states were often productive grazing lands, but Malay lands away from rivers were of little economic use. There was less incentive to mark off land deemed unproductive. Indeed, Greek colonies in northern Africa placed markers along the coast to separate neighboring territories but, seeing little economic potential in the desert interior, made little effort to extend borders inland from these coastal markers. A similar pattern manifested in the seventeenth and eighteenth centuries, as Europeans marked the coasts of Southeast Asia but left the interior jungles without demarcation. The Greeks were also content to leave neutral zones between territories, but the Mayans appear to have assigned all their territory to one city-state or another. This might be explained by the Greeks' need for grazing lands, which could be used in common with neighboring city-states, while the Maya practiced intensive agriculture, which required clearer partitions of land. In contrast to both, the Malay city-states relied more on the sea and rivers for their food, as well as for trade. These different environmental, agricultural, and economic activities fostered the development of differing territorial strategies. Despite the apparent differences in their attitudes toward borders, the Greek *polis*, the Mayan *ahawlel*, and the Malay *negeri* all represent territorial practices that harnessed the manpower, resources, and production in their hinterlands for the benefit of the urban center.

Empires

The empire was the third political-territorial structure common in antiquity. Empires usually formed when one city-state, or less often a nomadic group, succeeded in establishing political and economic dominance over an expansive, multiethnic territory beyond its immediate, traditional homeland. Most empires were governed by hereditary monarchs whose legitimacy was supported by some notion of universal kingship and a broader effort to "civilize" other peoples. This ideology often had a religious basis; that is, the monarch claimed a divine mandate to govern. As a result, empires tended to pursue policies aimed at subjugating neighbors, whether they were nomadic groups, city-states, or other empires. It is easy to think of empires as simply successful city-states that grew very large, similar to the transformation of the Babylonian or Roman city-states into their respective empires. Yet empires required more expansive and hierarchical territorial structures to govern their varied peoples and regions effectively.

In general, early empires can be grouped into two broad categories based on whether territorial control was exercised through direct or indirect means. Some empires developed relatively centralized systems that focused policymaking authority in the imperial capital. The territory beyond the imperial core was divided into provinces headed by regional governors. These governors, normally drawn from aristocratic families based in the imperial core, were responsible for maintaining law and order, implementing policies from the central government, and perhaps most importantly, overseeing the collection of taxes and resources to be shipped to the imperial capital. The Persian Empire, for example, was divided into approximately two dozen provinces each headed by a governor, or *satrap*, who ruled on behalf of the emperor. Similar arrangements have been documented in the Egyptian, Roman, and Inca empires.

Other empires relied on less centralized means of control. The Aztec Empire was essentially a city-state, Tenochtitlan, that succeeded in dominating the other city-states in central Mexico. Aztec rulers did not directly annex these territories; rather they generally allowed local kings to remain in place so long as they provided military service and sufficient tribute to the central government. Local kings retained considerable authority over local matters and their positions were hereditary, unlike the governorships in more centralized empires. Although they may appear quite different, these forms of rule shared a great deal. They were all intended to channel tribute and resources from the provinces or subject kingdoms to support and benefit the imperial government. The possibility that local kings or governors might try to break free was also a constant problem for both and often a more serious threat to the survival of the empire than hostile neighbors.

Regardless of their internal administration, empires sought some way to structure their territory and mark its external boundaries. The Roman Empire appears to embody the ideal of a highly centralized empire with clear external borders. The surviving portions of Rome's border defenses reinforce this perception for modern viewers. Remnants like Hadrian's Wall, an extensive network of forts, towers, and walls in northern England, suggest clear and definite limits of Roman territorial control. Yet the impression of Roman soldiers standing guard along the wall against barbarians on the other side is misleading. Rather than marking the limit of Roman control, the Romans intended to use Hadrian's Wall to project their authority well to the north. In this sense, the Roman use of Hadrian's Wall is not unlike the Chinese approach to its Great Wall. Both of these frontier fortifications suggest clearly defined limits to imperial authority but rarely marked an exact border between their respective empires and the northern tribal groups with which they both traded and clashed. Archeological evidence of settlement patterns suggests that pastoralists and sedentary farmers were commonly found on both sides of China's Great Wall, much like material remnants of Celtic peoples and Roman settlers span Hadrian's "barrier." In practice, these walls provided imperial troops a platform for controlling territory and regulating the movement of people and goods on both sides. The ability of Chinese and Roman forces to actually achieve these goals varied over time. The fact that successive Chinese dynasties were compelled to construct new walls over so many centuries and that Rome eventually abandoned Hadrian's Wall reflects the challenge of maintaining stable frontiers.

Although impressive structures like Hadrian's Wall or the Great Wall suggest clear linear borders, it is more helpful to think of early empires as possessing relatively fluid, indeterminate territories. Recent research on the Assyrian Empire, for example, highlighted the degree to which premodern states developed different territorial strategies for different sections of their frontiers. Assyrian rulers appeared to adapt frontier strategies based on an area's relative economic or strategic importance. In areas deemed important, the Assyrians took direct control by establishing new administrative centers. The forced importation

3. Ancient fortifications, such as the Great Wall of China, rarely marked a rigid political border.

of agricultural colonists to newly conquered regions increased the area's productivity and ensured a more loyal populace. In areas deemed less important, the Assyrians seemed content to maintain neutral zones occupied by village-level chiefdoms, which posed no security threat. It also appears that Assyrian control in its outer provinces was patchy, with imperial control clearly evident in some areas but largely absent in others. Instead of a contiguous territory, the Assyrian Empire is better described as a network of dispersed pockets of imperial control interspersed with areas lacking significant political organization. Similar variation has been found in the frontier policies of the Aztec and Sassanid empires. Depending on differences in the natural terrain, strength of opposing forces, and economic importance, some frontiers were administered indirectly through local rulers who pledged loyalty and tribute to the empire, other provinces were placed under direct imperial control, and still other stretches of frontier were basically left open.

Ancient polities as flexible territorial structures

Early nomadic bands, city-states, and empires employed a range of beliefs and practices for organizing and bordering territory. Given differing economic, cultural, political, and environmental contexts, it should not be surprising that groups in different parts of the world and in different times developed varied approaches to territory, borders, and governance. These approaches were more nuanced and flexible than one might imagine. Several scholars have argued that these ancient governments were more focused on controlling the movement of people than controlling actual territory since labor supply was the key factor in maximizing the agricultural production necessary to support the ruling class. Although still a matter of debate, it appears that the various territorial and border strategies were as much about ensuring a steady and reliable labor force as laying claim over territory. If correct, this helps explain why many of these ancient governments were generally satisfied with relatively vague borders.

However, modern maps tend to depict these early nomadic groups as nonterritorial, city-states as mere dots, and empires as clearly bounded political territories. This cartographic schema has less to do with the realities of the time or historical evidence than with the projection of contemporary assumptions of the modern world back on to previous centuries. Such a simplistic perspective leads to two general misconceptions of state formation, territoriality, and borders in antiquity. The first interprets history as an inexorable evolution from nomads to city-states to empires, or from small and simple to large and complex. Yet there are numerous examples of empires that fractured into smaller-scale polities or nomadic groups, which built empires without passing through the city-state "stage" and which refute such a linear narrative. A second misconception presumes that ancient polities possessed uniform control over territories marked by clear borders. Although early nomadic tribes, city-states, and empires may have claimed absolute authority over a given area, a variety of strategies for political-territorial control were used over the centuries that provided governments varying degrees of authority and integration within their nominal territories.

Internally, these ancient governments may have functioned more like flexible, patchwork networks of territorial control interlinked by strategic transit corridors. Externally, territorial borders were more likely to feature transitional frontier zones containing a fluid mixture of imperial strongholds, client states, opposing forces, and areas of indeterminacy, instead of sharp lines separating clear spaces of sovereignty. Areas with limited accessibility, either through sheer distance from the capital or natural obstacles like mountains, often straddled these distinctions between internal and external and provided refuge for various "tribal" groups to operate autonomously, despite occupying territory nominally claimed by some higher sovereign. The desire to eliminate these territorial gaps in the exercise of political power provided a powerful impetus for new territorial and bordering strategies that ultimately contributed to the formation of the modern state system.

Chapter 3
The modern state system

In the ancient world, territorial control did not always rely on clearly defined borders. Yet the variable and flexible approaches applied in the past gradually gave way to more standardized and rigid notions of borders, territory, and sovereignty within what is generally known as the modern state system. Referring to the division of the Earth into political entities mutually recognized as possessing absolute sovereignty over the populations and resources within a given territory, the organization of global space into independent states is largely taken for granted today. Yet the development of this system was a gradual and, in many ways, an incomplete process. The modern state system first emerged in Europe and spread to the rest of the world through colonialism. Understanding the emergence of the state system provides a foundation for understanding current debates about the role of borders and territory in our increasingly globalized world.

The origins of modern states

The modern state system traces its origins to complex economic, social, and political changes occurring in western Europe beginning as early as the eleventh century. Although some familiar names, like France and England, could be found on Europe's map at this time, these entities were structured quite differently than their modern successors. Instead of distinct sovereign states, medieval Europe was organized around what became known as the feudal system. Feudalism was a form of political organization involving a complex system of privileges and responsibilities between lord and vassal. As the Carolingian Empire declined during the ninth century, Frankish rulers found it increasingly difficult to pay their military subordinates. Instead of monetary payment, the kings appointed their commanders as vassals with the right to use a portion of the king's land and control its economic production. In return, the vassals promised loyalty and military service to the king. As a result, feudalism was based more on personal oaths and obligations than on rigid territorial organization. Kings and vassals were bound by their personal commitments regardless of their territorial location.

Feudalism seemed to offer a clear hierarchy, with lesser nobles at the bottom and kings at the top. Indeed, the king retained ultimate ownership of the territory and could reassign it if the vassal was disloyal or died. However, as central authority weakened even further, vassals were able to gain hereditary title to their lands. Gradually, local nobles became largely independent and appointed their own vassals as kings lost much of their control over their vassals' territories. Marriages and land transactions between noble families, elaborate inheritance customs, and military conquest further complicated the situation. The result was a system of decentralized political authority, scattered territories, and overlapping jurisdictions. While the dukes of Burgundy, for example, possessed large land holdings within the kingdom of France, they also held considerable territories in the Holy Roman Empire. On top of this, the Roman Catholic Church claimed a degree of universal authority. Therefore, the dukes owed some form of allegiance to the French kings, German emperors, and Catholic popes simultaneously. Given this confused structure, precise territorial borders were not necessarily needed or helpful so long as taxes were collected, services rendered, and oaths fulfilled, especially in sparsely populated areas.

Ironically, a political structure based on state sovereignty within clearly marked territories has its roots within a feudal system characterized by overlapping jurisdictions and vague borders. Scholars debate the exact causes and timeline, but there is general agreement that the modern state system began to form by the late Middle Ages as centralized governments exercised increasing political and economic control over defined territories. The reasons for this shift are complex but generally involve a series of economic changes beginning around 1000 CE, including increased agricultural yields, specialized production of goods, monetary exchange, and long-distance trade. These changes fueled the development of cities and an increasingly powerful new class of urban craftsmen and merchants.

The growing power of cities disrupted established feudal relationships and provided opportunities for new political arrangements. Cities generally shared the same objectives of political security and stability that would foster economic growth and trade. Kings and feudal nobles hoped to harness the growing wealth of these cities in their struggles with each other. Yet differences in the relative strength of cities, nobles, and monarchs across Europe led to different political outcomes, namely the emergence of city-leagues, city-states, and territorial states. This redistribution of power was accompanied by new approaches to territorial organization that gradually replaced the decentralized logic of feudalism and the universalistic claims of the popes and emperors.

In Germany, cities were not strong enough to act individually, and the German kings were weakened by their efforts to incorporate Italy into the Holy Roman Empire. As a result, cities banded together into city-leagues to defend their interests against the nobles. These leagues were confederations of scattered cities featuring decentralized decision-making, limited authority over members, and vague borders beyond their city walls. In Italy, some cities grew large enough to defend their economic and political interests individually. As a result, they resisted efforts to establish centralized authority under the pope or emperor and instead formed independent city-states. These city-states had external borders similar to territorial states. Yet their internal organization was often quite fractured. The dominant city-state ruled its surrounding villages and lesser cities in an exploitive manner, causing them to continually challenge the authority of the central government.

French cities were the weakest and could not defend themselves as city-states or even in city-leagues. The French kings of the Capetian line were also relatively weak and controlled a limited area. As a result, the French kings and urban merchant class banded together against their common threat, the nobles. The cities supported increased central control under the king, and in return, the king protected the cities against the nobles. The king and cities also had a shared interest in fostering favorable economic conditions, including regular systems of taxation, centralized bureaucracy, uniform legal systems, a standing military, transportation infrastructures, and reducing the influence of the nobility, church, and guilds. This cooperation between the French kings and cities led, by around 1300, to the emergence of a territorial state featuring relatively clear external borders and centralized internal sovereignty exercised by a monarch.

These three different forms of political organization co-existed as feudalism faded. Gradually, territorial states proved more efficient at ensuring the security and stability desired by political and economic elites. By 1500, territorial states ruled by hereditary monarchs were coalescing across western Europe. Indeed, backed by increasingly centralized administrations and improved military capabilities, some monarchs claimed absolute authority based on a divine right to rule. In many respects, the absolutist monarchy was a compromise that aimed at reconciling certain aspects of feudalism, namely monarchical rule and privilege, with the growing economic and political power of an urban-based middle class. The incompatibility of the feudal versus absolutist territorial systems eventually came to a head as the Reformation and wars of religion embroiled much of Europe in more than a century of bitter conflict.

The Peace of Augsburg in 1555 and the Peace of Westphalia in 1648 were meant to resolve these conflicts by outlining the basic principles for a new territorial-political order—the modern state system. After decades of intermittent conflict, European states agreed to recognize each other as possessing exclusive authority over specific territories. This had three main implications. The first was the notion of autonomous sovereignty, or the principle that states would be free to govern their territories without outside interference. Second, states would be regarded as the only institutions that could legitimately engage in international diplomacy and war. Finally, the claim and exercise of absolute sovereignty required monarchs to mark the precise borders of which territories, populations, and resources were included in their territories and which were excluded. Ambiguous frontiers may have been compatible with the flexible and overlapping nature of territorial control in medieval Europe, but they fit poorly with this new notion of absolute state sovereignty.

From natural borders to national borders

Supported by burgeoning capitalist economies, expanded state bureaucracies, advanced means of military organization, and the emergence of modern nationalism, European states gradually acquired the ability to demarcate the precise limits of their sovereign territories. This development was readily apparent in advances in surveying and cartography that allowed ever greater accuracy in measuring and demarcating space. Yet it was unclear what criteria should determine where these boundaries would be drawn. The idea that borders should follow physiographic features like rivers or mountain ranges predominated during the seventeenth and eighteenth centuries. Presuming that nature had already marked out the territorial extent of each monarch's authority offered "rationality" to setting boundaries. If "correct" limits for sovereignty were predetermined by nature, states simply had to locate them and adjust their borders accordingly.

While the idea of using natural borders seemed an objective basis for setting state borders, in practice individuals tended to select natural features that supported their geopolitical goals. French writers, for example, often argued that the Rhine River was France's natural eastern limit. This assertion just happened to coincide with French efforts to annex territories along the Rhine. Indeed, authors from France and other countries commonly selected natural features beyond the current borders of their state, while few argued that nature favored borders reducing the territory of their state. The idea of natural borders generally served as little more than a veiled excuse for territorial expansion.

Although claiming absolute power, monarchs faced increasing demands for greater political participation and representation. Control of the state gradually shifted to democratically elected governments as monarchs were either overthrown or reduced to figureheads. This was an uneven transformation involving many factors, but perhaps the most important was the emergence of nationalism as a mass phenomenon. The rise of modern nationalism transformed ideas about state sovereignty. If a nation is defined as a group of people who believe they constitute a unique grouping based on shared culture, language, history, and the like, then nationalism is a political ideology that assumes the nation commands the primary allegiance of its members and possesses an intrinsic right to self-determination within its own sovereign state. Previously, the monarch was the embodiment of the state and sovereignty. This idea was exemplified by the French king Louis XIV who, shortly before his death, allegedly remarked "I am the State" (*L'état c'est moi*). Following the American and French revolutions as well as the steady expansion of parliamentary rule in England, this idea of monarchical sovereignty was gradually replaced with popular or national sovereignty. The state came to embody the sovereignty of the nation, not the monarch. The idea of the nation-state, where the political borders of the state would coincide with the cultural boundaries of the nation, became the ideal, although not the norm, by the end of the nineteenth century.

The formation and diffusion of the idea of a German fatherland illustrates this shift from monarchical to national sovereignty. As aspirations for national sovereignty spread in the wake of the French Revolution, ethnic Germans found themselves fragmented between numerous principalities, bishoprics, and petty states. As a result, German nationalists called for the creation of a unified German nation-state, raising the obvious question as to which areas should be included. The German writer Ernst Moritz Arndt provided one of the most popular answers in his poem *The German Fatherland*. Arndt argued that the German fatherland stretched "Wherever is heard the German tongue, And German hymns to God are sung!" Rather than relying on natural features to mark this proposed state, Arndt and other German nationalists believed ethnolinguistic criteria should be paramount; that is, Germany should include all German-speakers. Although a German state was created in 1871, significant German-speaking populations were excluded. Expanding the state to include all ethnic Germans would remain a major objective for nationalist groups, including the National Socialist (Nazi) movement, with disastrous results.

As this new idea of state sovereignty gained acceptance, it became increasingly common to argue that borders should correspond to ethnic and linguistic divisions, instead of focusing on natural landmarks. So the borders of the state of France, for example, should be drawn to include all French populations and lands. Italians also pushed for a new state that would unify their nation, while numerous nationalist movements challenged the multinational empires of eastern Europe. What began as a state system based on monarchial sovereignty over a royal domain gradually evolved into a state system, or perhaps better described as a nation-state system, based on national sovereignty over an ethnic homeland.

Nation, state, nation-state

Like the previous notion of natural borders, the idea of drawing borders to achieve national sovereignty would prove difficult to realize, as the peace negotiations following World War I made clear. President Woodrow Wilson's Fourteen Points for ending the war and preventing future conflict reflected this new view that state borders should correspond to ethnolinguistic divisions. Among his proposals, Wilson called for redrawing the borders of Italy "along clearly recognizable lines of nationality," providing the Turkish people "secure sovereignty," and creating a new Polish state including "the territories inhabited by indisputably Polish populations" with "political and economic independence and territorial integrity." Finally, Wilson proposed an "association of nations . . . for the purpose of affording mutual guarantees of political independence and territorial integrity to great and small states alike."

Wilson's proposals repeated established Westphalian principles but also reflected the new language of nationalism. States were still regarded as independent entities possessing sovereignty over distinct territories, but their borders should reflect differences in nationality and language. Multinational states, like the Ottoman and Austrian empires, were seen as antiquated. These would be replaced by new nation-states; that is, a Poland would be created for the Poles, a Hungary for the Hungarians, and so forth. Yet reconciling state borders with nationality and language proved just as difficult, subjective, and contentious as attempts to set international borders based on natural features. Indeed, ethnolinguistic groups rarely have sharp dividing lines, so nationalists argued over which nation had the strongest claim to particular territories. Much of European political history since 1800 involves efforts to revise the region's feudal/absolutist borders to fit this new nation-state framework. Borders played a dual role in these processes as both causes and effects. In some ways, national or ethnic identities provided the cause for marking new state borders, but in other ways new state borders had the effect of furthering the creation of new national or ethnic identities. Unfortunately, these new bordering processes were marked by massive death, dislocation, and numerous territorial revisions, especially during and after World War II.

As a result, Europe evolved from a region of monarchical states to a region of democratic states whose borders generally correspond with the larger nationality groups. Poland during the interwar period, for example, was a multinational state with perhaps only 60 percent of the population identified as Polish. The brutality of World War II brought significant demographic and territorial adjustment so that Poland is now 96 percent Polish. The upheavals of the post–Cold War period marked a continuation of this trend as the Soviet Union, Czechoslovakia, and Yugoslavia split apart largely along ethnolinguistic lines. This is not to suggest that Europe's current borders correspond perfectly to national differences or that Europe's borders have been finalized. Indeed, significant minority populations remain across Europe, and there are several nationalist movements that advocate border changes to create additional nation-states (for example, the Scots in the United Kingdom, the Basque in Spain, or the Dutch in Belgium).

These continuing efforts to adjust borders and reallocate territory highlight an inherent tension within the foundation of our contemporary international system, namely the conflict between guarantees of state sovereignty and territorial integrity on one side and recognition of a national right for self-determination on the other. It is extremely difficult to reconcile the fuzzy cultural boundaries among ethnolinguistic groups and the sharp political borders dividing sovereign states. This contradiction is present in the founding charter of the United Nations. Established in

1945, the United Nations was intended to further international cooperation, especially the goal of ending war. Article 1 of the charter outlines the purposes of the United Nations, including promoting "friendly relations among *nations* based on respect for the principle of equal rights and self-determination of *peoples*" and serving as a "centre for harmonizing the action of *nations* in the attainment of these common ends." But then Article 3 limits membership to "peace-loving *states*," while Article 2 recognizes the "principle of the sovereign equality" for all member states and prohibits the United Nations from interfering in issues "within the domestic jurisdiction of any *state*" (italics added). So despite its name, the United Nations is not an organization of nations but rather of states. The states of Turkey and China are recognized as equal, sovereign entities, but the Kurdish or Tibetan nations are not. This basic contradiction is a primary cause for much contemporary inter- and intrastate conflict.

Colonialism and sovereignty

During this transformation from feudal to absolutist to territorial nation-states in Europe, Europeans were simultaneously engaged in colonial expansion. Colonialism refers to a process whereby a state establishes direct political and economic control over territories beyond its commonly recognized borders. European colonialism was motivated by diverse factors including economic opportunism, geopolitical rivalries, missionary passions, and settlers seeking better opportunities. Regardless of the exact motivations, the result was usually an unequal relationship that benefited the colonial power and its settlers more than the colonial territory and its indigenous inhabitants. The establishment of European colonial control over much of Africa, Asia, and the Americas brought dramatic and often destructive changes to colonial lands, societies, and economies, including the imposition of European norms of sovereignty, territoriality, and borders. Although these non-European societies possessed their own conceptions of territorial organization, it was largely the political and geographical notions championed by Europeans and exported through colonialism that provided the basis for the modern state system.

The expansion of European forms of territorial organization was integral to colonialism. As one of the first steps in establishing sovereignty over these new territories, European states began mapping and reorganizing them to conform to the territorial state model that was emerging back in Europe. From the European perspective, colonial territories were basically "empty" lands to be claimed despite the obvious presence of established populations, societies, and governments. Preexisting systems of land ownership and resource access would be radically transformed or obliterated.

The Treaty of Tordesillas in 1494 was one of the first attempts to impose European notions of territorial sovereignty beyond Europe. Spain and Portugal, initially the leading colonial powers, hoped to divide all non-European territories among themselves and prevent claims by other European countries. The treaty designated a meridian running roughly through the middle of the northern Atlantic Ocean and South America. Spain claimed all lands to the west of that meridian, while Portuguese claims would extend to the east. The Treaty of Zaragoza in 1529 created another meridian that divided Spanish and Portuguese claims in the Pacific Ocean on the other side of the world. These treaties had little impact since other European powers rejected them immediately. Nonetheless, they marked the beginnings of a long-term process in which the European territorial state became the global norm.

Colonialism was an uneven process in terms of the extent and effectiveness of European control. In general, European colonialism can be divided into two phases. The first phase began in the late fifteenth century. Because Europe's emerging territorial states often lacked the capacities to finance large overseas operations, many colonial efforts began as semiprivate commercial ventures. These chartered companies used private investment to fund the exploration, acquisition, and administration of colonial territories. In return, European governments would grant these companies special economic privileges, such as trade monopolies with the colony, to generate profits for the investors. This commercial colonialism was a rather indirect form of control in which company officials were the effective rulers of the colonies, although they remained nominally under the authority of a government back in Europe.

Some of these chartered companies, like the British East India Company, became quite powerful and controlled military forces that rivaled some European states. Nevertheless, company officials controlled relatively modest territories, usually small coastal settlements where they conducted trade with indigenous merchants and rulers. These companies were not primarily interested in controlling territory. Rather, they were interested in controlling the flow of commodities between the colony and Europe. As a result, they were largely unconcerned with establishing vast territorial claims or clear borders. These were unnecessary expenses, so long as trade goods continued to flow and rival companies kept their distance.

Despite the focus on controlling the movement of goods, chartered companies gradually found themselves administering ever greater territories, often as an ad hoc preemptive strategy to block rival claims. Chartered companies were soon overwhelmed by the burdens of government. At the same time, states in Europe were increasing their abilities to exercise effective territorial control. This marked the beginnings of a second stage in European colonialism as earlier forms of indirect rule gave way to European states assuming direct and formal control over colonial administration. This transition occurred earlier in the Americas than in Africa and Asia, but by 1900 European states had gradually extended claims of direct sovereignty over most of their colonial territories.

Creating colonial borders

The shift from commercial to state colonialism greatly expanded the territory controlled by European states. This transition also entailed the extension of the European territorial state model to colonial territories and subsequently the need to create clear borders. This was perhaps most evident during the so-called Scramble for Africa. After assuming more direct control from the coastally oriented charter companies during the nineteenth century, European states quickly moved to claim interior parts of Africa. There seemed a real danger that competition could lead to war, so European leaders met at the Berlin Conference in 1884/85 to partition Africa. It was through this conference and later negotiations that Europe's leaders largely created Africa's modern political borders. They did so with limited information about the lands and peoples they were reorganizing and without input from Africans. A similar process was under way across much of Asia. Although most of the world's political entities were bounded by relatively vague and ambiguous frontiers at the beginning of the sixteenth century, the situation had changed dramatically by 1900 as colonial powers hurried to mark the exact limits of their territorial claims.

European colonial expansion and border demarcation even impacted areas that were never brought under European sovereign control. During the nineteenth century, Britain and France were expanding their colonial possessions and projecting European territorial norms in Southeast Asia until only the kingdom of Siam (roughly modern-day Thailand) remained independent. The Europeans were operating under the assumption of absolute sovereignty, but the Siamese kingdom was structured around a form of shared sovereignty between the central monarch and local rulers in outlying areas. British envoys repeatedly contacted the Siamese rulers to negotiate a border between their territories, but for the Siamese monarchs, external borders mattered little so long as the local leaders remained loyal and served the central

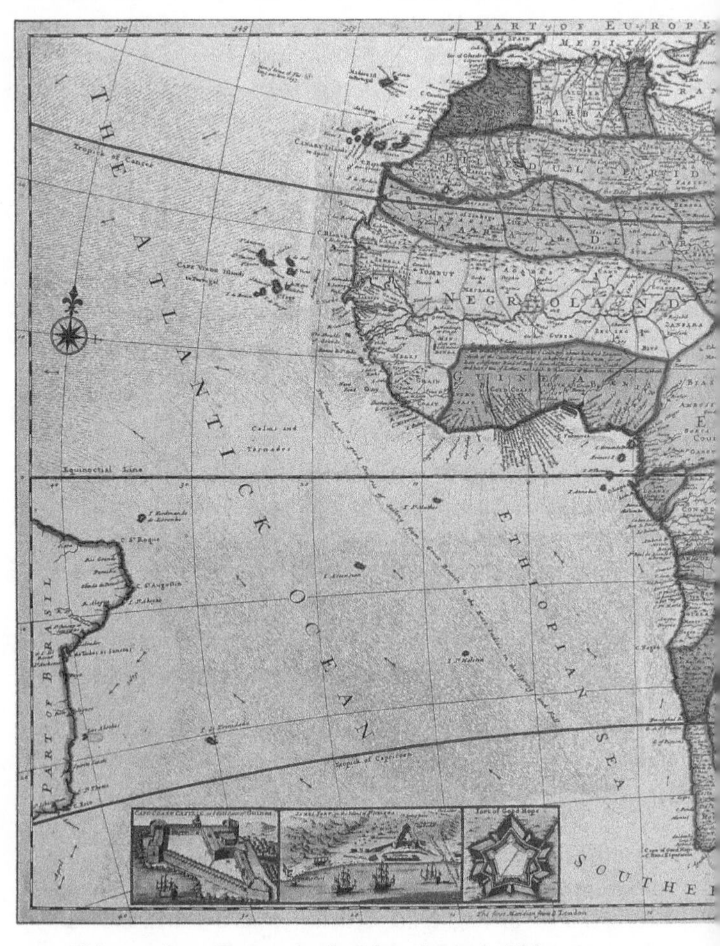

4. **This map shows European colonial territories in Africa, 1710.**

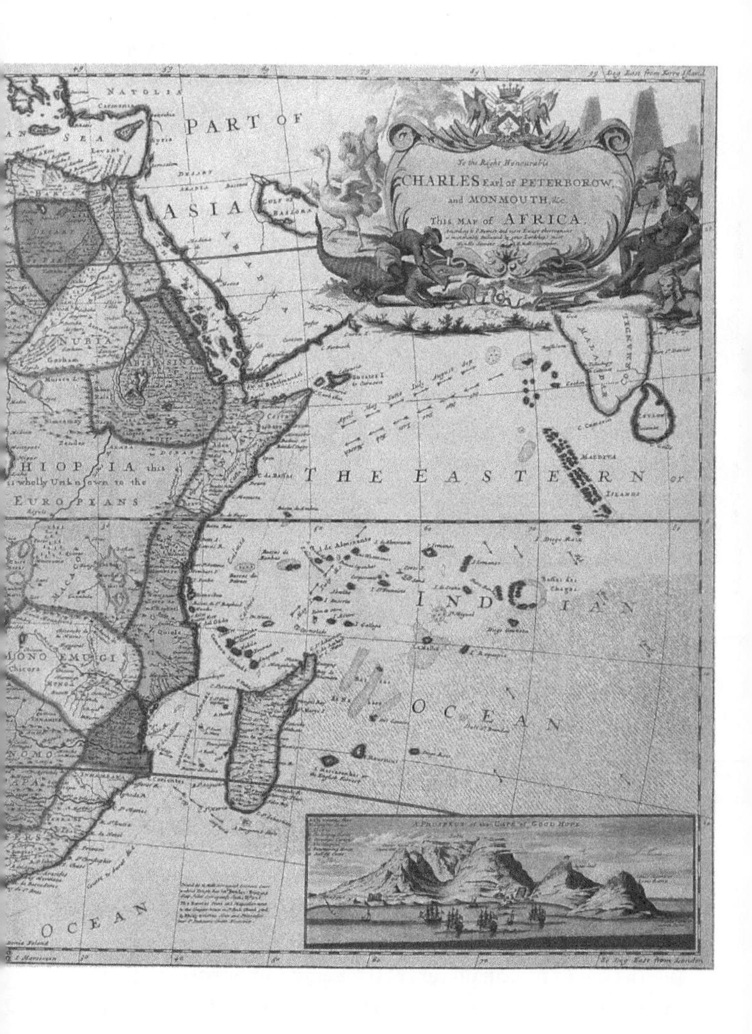

NATOLIA
PART OF
ASIA
To the Right Honourable
CHARLES Earl of PETERBOROW,
and MONMOUTH, &c.
This MAP of AFRICA
NUBIA
HIOP IA this
is wholly Unknown to the
EUROPEANS
THE EASTERN or
INDIAN
OCEAN
MONO EMUGI
OCEAN

government. After continuous European encroachment into their territory, the Siamese government adopted Western notions of borders and negotiated clear boundaries for their sovereign territory. This created a "geo-body," or territorial outline of a homeland, that fostered the formation of a Thai state and nation.

Westerners generally saw colonialism as "civilization" advancing and bringing various benefits to colonial subjects, including new ideas of territory, sovereignty, and governance. This also provided a convenient justification for conquering more land. It also ignored how uneven and incomplete colonial control was in many areas. The expansion of European control usually depended on the cooperation or co-optation of some local elites. Even after the British state assumed direct control of colonial administration over British India in 1858, for example, sovereignty remained confused. Even though the British government exercised nominal sovereignty, hundreds of Indian princes remained largely autonomous concerning domestic affairs within their territories, such as taxation and law enforcement. As a result, British officials organized several surveys to mark the boundaries between British and princely jurisdictions. Although world maps generally depict colonies with solid colors suggesting political uniformity, arrangements of overlapping or shared sovereignties were quite common.

Colonial rule may have been incomplete and short-lived, but the imposition of the territorial state model would have far-reaching implications. Indeed, the origins of most contemporary international borders outside of Europe have some colonial connection. Some originated from agreements between competing states as they sought to establish initial claims of sovereignty, such as the European powers at the Berlin Conference. In these cases, the borders between respective colonial territories, such as between British and French colonies, later became the new international borders between independent sovereign states, especially during the wave of decolonization following World War

II. Other international borders emerged when larger colonial territories were partitioned along former internal administrative boundaries, such as French Indo-China or British mandates in the Middle East.

The British withdrawal from India in 1947 provided one of the most striking examples of the transition from colonial rule to independence. The external borders of British India were set, but disagreements between Hindus and Muslims precluded a simple transfer of sovereignty. As a result, the British government decided to partition British India into two states: India, which would include predominantly Hindu areas, and Pakistan, which included two predominantly Muslim areas separated by nearly 900 miles (1,450 kilometers). It was impossible to draw lines perfectly dividing Muslim and Hindu communities into their respective states, so millions of people found themselves on the "wrong" side of the new border. Many people migrated to their "correct" state, but tragically, thousands were killed in the ensuing confusion and violence. The problems of partition in South Asia were further complicated by the semisovereign Princely States. These princes were allowed to choose which state to join. Most Hindu princes ruled majority Hindu populations and joined India, while Muslim princes joined Pakistan. The Kashmir region had a majority Muslim population but a Hindu prince who joined India. This led to war between India and Pakistan, and the division of the area. Kashmir's status remains unsettled, and with China claiming a portion of the territory, this border dispute is among the most volatile in the world. Indeed, post-partition Pakistan proved untenable, and East Pakistan broke away as the independent state of Bangladesh following a bloody civil war in 1971.

Given that the number of sovereign states in the world increased from approximately 70 in 1945 to more than 170 by the 1980s, it might seem surprising that colonial-era borders were retained with relatively few modifications. This again highlights the degree to which the modern state system prioritizes territorial integrity of states over claims of national self-determination. States have been extremely resistant to the redrawing of international borders, fearing it could ultimately undermine their own territorial integrity. The partition of British India also served as a warning that redrawing borders would likely trigger disagreements and possibly violence. Although most contemporary international borders are relatively recent creations, and quite arbitrary ones at that, they have assumed a degree of permanence and timelessness. This air of permanence is, however, far less applicable to substate borders.

Substate borders and boundaries

In addition to their external borders, states also contain numerous boundaries that differentiate domestic spaces. Some of these are official administrative or bureaucratic jurisdictions; others mark unofficial or informal social groupings. Official administrative borders are generally easily identifiable, and most people are aware of living within some lower-level jurisdiction. Despite this familiarity, substate borders vary greatly from state to state in terms of their function, impact, shape, purpose, and even basic terminology (for example cantons, departments, states, republics, provinces, oblasts, aimags). These variations reflect differing assumptions concerning the exercise of sovereignty, as well as practical considerations like population density, cultural differences, and natural features.

The form and function of the fifty U.S. states offers an interesting example. The formation of the original thirteen colonies reflected varied political considerations of the British monarch when granting colonial charters. As a result, the territories and populations of the colonies varied significantly. In response, the U.S. Constitution ensured some measure of equal representation among the states by creating a senate with two senators per state. The U.S. government also endeavored to create roughly equal-sized states based on geometric ideals in its newer western territories. The result is that eastern states feature irregular shapes and greater variation in size compared with western states.

The United States also adopted a federal system of government, meaning that sovereignty is shared between the central government and member states. The U.S. Constitution, for example, limits the power of the central government to certain functions like diplomacy and defense, while states retain sovereignty over other functions like education and law enforcement. Many other countries have adopted federal systems granting substate governments varying degrees of responsibility, authority, and representation. In contrast to federalism, unitary states feature the concentration of sovereignty in a single governmental unit. In unitary states, substate entities merely enact policies emanating from the central government, and these lower-level jurisdictions have little or no authority to make independent policy decisions.

The desire to achieve equality among the states meant that the United States drew many of its internal political borders as though operating on a blank canvas. In contrast, many intrastate borders around the world reflect ethnic or tribal communities, historic realms of feudal lords, estates of princes, or colonial possessions. Even in diminished form, these borders have endured and play powerful roles in shaping political participation, the provision of government services, and cultural identity. This is not to say that borders and the territories they demarcate are permanent—historic atlases provide numerous examples of defunct states and borders. The point, however, is that every border has a story. Every line on a map, every marker in the landscape, was derived from some complex negotiation of power and culture.

The formation of substate borders often creates significant spatial differences in educational opportunities, political representation, government services, or financial services. Unfortunately, the processes creating these borders are highly susceptible to political manipulation. "Gerrymandering" refers to the drawing of substate borders with the intent to advantage a particular group. Legislative districts in the United States, for example, are commonly gerrymandered to make sure that a specific political, racial, or social group constitutes a majority and is therefore likely to win any election. Censuses that collect economic, racial, and cultural data often provide the basis for these politically motivated borderings. There are many similar examples around the world. The division of Israeli and Palestinian territories resembles an archipelago where ethnoreligious segregation has de jure sanction, while several post-Soviet states have recently revised electoral districts to reduce the political influence of their Russian minority populations.

These geographies of difference extend beyond official demarcations to include various informal social, economic, or cultural boundaries. These reflect and reinforce social categories of wealth, power, and privilege in both developed and developing countries. Residential patterns in American cities are closely tied to economic and racial differences with lower income minority groups often concentrated in inner-city neighborhoods. Similar patterns of residential segregation are evident in European cities with lower-income immigrant groups clustered in specific districts. These examples highlight complex economic, political, and social forces that create informal spatial restrictions on residential choices for minority and low-income groups in developed countries. Similar processes are evident in the developing world as well. Communities of privileged retirees from developed countries have emerged in parts of Central America and the Caribbean. These residential borders help establish and reinforce social differences, as well as setting parameters for daily mobility, access, and interaction.

The growth of gated communities is one of the most recent developments in residential segregation. As the term suggests, these neighborhoods utilize walls, fences, checkpoints, guards, and surveillance equipment to create spaces of security and privilege. Gated communities were initially designed for wealthy households in developed countries, but they have become increasingly common in other regions. For example, gated communities in developing countries often contain high concentrations of foreign nationals. These range from the previously mentioned retirement communities in Central America and the Caribbean to residential compounds for expatriate workers in Saudi Arabia and other Gulf states to the nouveau riche of the postsocialist realm.

Lower-income neighborhoods exhibit similar tendencies toward territoriality, although they lack the security infrastructure and official recognition. Gang territories, for example, are regions relying on unofficial demarcations of territorial control. Gang members maintain some measure of privilege for themselves within the territory they control. Although graffiti and the display of "colors" may seem as simple markers of lower socioeconomic status, they can actually represent the territorial claims of particular groups.

In a more benign sense, street banners, murals, and commercial signage can also demarcate zones of privilege or belonging for certain people. This can take the form of ethnic neighborhoods,

5. This gated community in California creates a separate residential space for the wealthy.

religious regions, linguistic communities, and gay/lesbian areas. Although lacking official sanction by the state and rarely found on official maps, these borders are based on the same principles of territoriality as international borders. They are designed to distinguish spaces controlled by members of a specific social group, symbolize the limiting of access to others, and compel certain norms of behavior. The numerous "Chinatowns" across North America, Asia, and Europe, for example, often feature arched entranceways and bilingual signs that associate these specific neighborhoods with a particular ethnic community. These types of ethnic enclaves are common features of cities with large immigrant populations and often show up on city maps and in guidebooks such as "Little India" in Singapore or "Japantown" in São Paulo.

The idea of globalization suggests a gradual detachment of certain identities from particular regions, threatening long-standing connections between people and place. Yet new international linkages present as many possibilities for "re-territorialization," as "de-territorialization." Indeed, many groups, ranging from ethnic communities to socioeconomic classes, are likely to respond to these broader changes by cordoning themselves off within socially or culturally homogeneous areas. In this context, substate administrative borders and informal social boundaries provide a means for negotiating new forms of cultural, political, or socioeconomic belonging. The prospect for significant re-bordering within states also coincides with new realities for borders between states.

Chapter 4
The practice of bordering

Territorialization is a twofold process involving the division of land between social entities and assigning specific meanings to the resultant places. Borders are integral to territorialization in their ability to symbolically perpetuate meaning and physically shape the mobilities of human beings. Rather than the sharply bounded territorial containers associated with modern nation-states, the nature of borders and practices of territoriality have clearly fluctuated over time. This perspective provides a foundation for considering alternative dimensions of boundaries, bounded space, and the processes of "bordering" in the contemporary world.

Traditionally, research has focused on the physical aspects of borders, like walls, fences, and observation towers. More recently, scholarly attention has shifted to understanding borders and bordering as processes, rather than things. The word "border," in essence, has become as much a verb as a noun. This shifting focus has given rise to many new questions: Do good borders make good neighbors or are borders impediments to international cooperation? Will state sovereignty continue to be prioritized over national independence? Will nation-states remain the primary global political/economic actors? Is a borderless world possible or desirable? These and other questions are pertinent to the evolving nature and role of borders in the twenty-first century.

International borders in transition

The unprecedented geopolitical changes of the late 1980s and 1990s, most notably the collapse of the Soviet Union, the fall of the Berlin Wall, and the proliferation of regional trade blocks (EU, NAFTA, ASEAN, etc.) compelled a broad reassessment of border functions, practices, and meanings. The rigid bipolar structure of the Cold War was collapsing, but it was unclear what would replace it. Speaking before the U.S. Congress in 1990, President George H. W. Bush proclaimed that "a new world order" was emerging based on shared values of freedom, justice, and peace, which would foster greater international cooperation. Many disagreed with President Bush's optimism, but most experts concurred that the international system was entering a transformative period and hotly debated the nature and direction of this evolution.

Much of the discussion focused on the idea of globalization, which refers to a set of political, economic, social, and environmental processes, which span international borders. Many scholars, politicians, business leaders, and activists believe that globalization's promotion of interaction and integration is rapidly eroding the importance of nation-states and their borders. Approaching this topic from a business, communications, or information technology perspective, some predict globalization will result in increased economic efficiency, technological diffusion, and overall rising standards of living. These rather optimistic views are countered by fears of economic decline in developed countries as companies shift jobs overseas or unrelenting waves of impoverished immigrants leading to increased cultural heterogeneity and competition for economic opportunities. Despite their differences, both scenarios envision a new world order with increasingly porous, permeable, and irrelevant international borders.

The terrorist attacks of the early 2000s have fueled a countertrend. Instead of increasing border permeability, many political leaders around the world and their constituents have sought greater control over their external borders. In addition to enhanced searches at airports, international travelers face stricter visa requirements, document examinations, and immigration procedures. Even the borders themselves are changing as new walls, fences, and observation towers seem to emerge daily.

The international system exhibits contradictory trends as globalization appears to entail both threats and opportunities. Although globalization is seen by many as a direct challenge to the nation-state system, state borders and territorial sovereignty have never been consistently enacted, performed, or perceived. Even as the nation-state system formed, both security and opportunity were part and parcel of broad-ranging global systems such as colonialism, imperialism, and mercantile capitalism. Although these systems offer precedents for the expansion of human connectivity and global interrelatedness, the intensity of contemporary interaction and interdependency among provinces, states, and global regions is unique. This has obvious implications for the fundamental role of territory and processes of bordering.

Globalization and territory

Many people today are questioning the primacy of state territorial sovereignty. Indeed, it is unclear if nation-states will remain the dominant actors on the international stage or might be gradually displaced by disparate international and substate actors. Many economists, business leaders, and journalists predict what may be termed "strong globalization." Building on earlier claims of the end of history or the end of geography, this perspective envisions a process of de-territorialization that collapses the nation-state system as borders become meaningless. In fact, Thomas Friedman's *The World is Flat* (2005) is based on the premise that globalization will allow all places to compete effectively in the global economy.

The prospect of "weak globalization" is more common among geographers, other social scientists, and international lawyers. These scholars emphasize the long-term variance and reversal in border creation and control. Some borders continue to function mainly as barriers, while other borders are being transformed into permeable sites of interaction, exchange, and cooperation. Harm de Blij's *The Power of Place* (2009), for example, highlights how standards of living for the vast majority of people correspond closely to the level of development within their country of birth. This approach recognizes the continuing capacity of borders to shape contemporary economies, societies, and politics as both barriers and bridges.

As the world's land area was divided into sovereign states, territorial control became a zero-sum scenario. A state can only gain if another state loses. This has been and will almost certainly remain a prime source of conflict. Indeed, much of international law aims at preventing or resolving territorial disputes. Yet it is important to recognize that international law developed within the overall framework of the modern nation-state system. Returning to the metaphor of the "territorial trap," international law takes the link between state, sovereignty, and territory as a given, and unsurprisingly almost always supports the status quo regarding international borders.

In fact, the primacy of a state's territorial integrity is explicitly codified by most international organizations. The United Nations Charter (Art. 2, Para. 4), for example, holds that "all Members shall refrain in their international relations from the threat or use of force against the territorial integrity or political independence of any state." Similar language is present in the founding charters of the Arab League (1945), the Organization of Islamic Cooperation (1969), and the African Union (2000). Some international organizations, like NATO and the European Union, require potential members to settle any external border disputes before joining.

This privileging of border stability ignores the fact that the formation of modern international borders was a rather uneven and arbitrary process that often divided national groups between multiple states or assembled multiple national groups within a single state. This basic disconnect between the cultural borders of nations and the political borders of states has contributed to many instances of organized violence over the past two centuries and continues to fuel many contemporary conflicts. Simply changing state borders to better reflect national differences seems an obvious solution. Yet there are at least three major obstacles. First, national groups often have overlapping senses of territoriality, meaning two or more nations claim the same land as their ethnic homeland. Second, the emotional element of nationalism raises the risk that groups will resort to confrontation, rather than compromise, in settling their claims. Finally, states are reluctant to undermine the sovereignty of other states fearing their own sovereignty could be undermined in response. This quid pro quo dynamic derives from the fact that state sovereignty is based on mutual recognition. Since the period of decolonization, the international community has generally recognized new states or significant border changes only in response to extraordinary events, like the collapse of Yugoslavia and the Soviet Union or decades of violence as in Eritrea and South Sudan.

Despite predictions of a borderless world, much of our daily lives, from human rights and national identity to natural resources and standards of living, remains fundamentally and inescapably linked to territory. Therefore, the prospect of political, cultural, or economic power completely detaching from territorial influence is unlikely. Indeed, states' interest in retaining and securing territorial sovereignty has increased since 2001. Rather than moving in a single direction, borders are being transformed in meaning and function by the contradictory pressures of global economic exchange (integrationism) and fears created by global security issues, especially terrorism (neo-isolationism). In essence, borders still matter but are taking on new roles and are understood to have effects more broad ranging than previously considered. Security, nevertheless, remains among the most prominent roles of borders.

Borders and security

Borders provide a powerful symbolic and practical means of dividing "us" and "ours" from "them" and "theirs." Indeed, the concept of national security and the state's capacity (some would argue obligation) to use force to defend its interests often centers on the maintenance of borders. Traditional perspectives on national security have therefore conferred a number of roles on borders and borderlands (regions directly adjacent to borders). These include serving as sites for evaluating neighboring states in terms of strengths, weaknesses, and intentions, as well as providing buffer zones protecting the state's core should a threat materialize.

Borders and borderlands also function as symbols of state securitization. They are often topical foci (sometimes scapegoats) for problems ranging from cultural change to economic downturns to any number of other social ills. In short, the border—those living near it, those crossing it, and its very image—helps shape political attitudes and policies. This latter point is worthy of expansion, as the concept of security is regarded today as far broader in scope than previously.

The role of state borders evolves with the nature of threats perceived. As noted earlier, the collapse of the Soviet Union was followed by a period of seemingly relaxed border enforcement in many parts of the world. Conversely, stricter border controls were regarded as remedies for a range of problems during the early 2000s, such as illegal migration, international terrorism, and smuggling. Ironically, most of the new threats are largely impervious to new border fortifications or patrols. For example, it is estimated that a mere 5 to 10 percent of drugs trafficked across international borders are interdicted. Furthermore, the impacts of natural disasters and disease are clearly not conditioned by political boundaries or their barrier infrastructures.

Given these new transborder realities, many experts regard traditional efforts at enhancing border security by reducing permeability as counterproductive. Current efforts focus on impeding the movement of people at the lines that divide one state from another, but a broader definition of "border space" may help authorities respond more effectively to new problems and threats. One novel approach to combating illegal migration might involve a de-securitization of border regions, which would allow for circular migrations, and greater cooperative efforts between states to facilitate economic development at the migrants' origins. For example, after the collapse of communism, Germany struggled with social tensions resulting from the return migration of ethnic Germans from the former Soviet Union. In response, the German government offered various incentives to this German diaspora community, such as sponsoring business ventures, to encourage them to remain abroad and not migrate to Germany. From this perspective, central governments must contend with the specific interests of border areas as well as the broader impact of border policies across their entire territories and those of neighboring and even distant states. This requires a more systemic and holistic approach to border security. Toward this end, new perceptions of borders as spaces, instead of simple lines on a map, focus attention on the array of new conditions, institutions, and actors influencing border permeability. In essence, borders are filters with a variety of different portals and levels of permeability.

Borders as permeable filters

As human constructs varying across space and time, borders and border permeability are inherently subjective phenomenon. Their specifics are shaped by legal, governmental, historical, social, political, and economic circumstances. Individual threats and pressures may increase or decrease the level of permeability for a given border at a given time. Sometimes permeability may even differ at various points along the same border.

The U.S.-Mexico border provides one interesting example. The presence of violent drug cartels in Ciudad Juarez, Mexico, triggered increased monitoring and restrictions on movement into and out of El Paso, Texas. In contrast, the border between Tijuana, Mexico, and San Diego, California, about 620 miles (1,000 kilometers) to the west, remains a relatively open portal for tourists and businesses. The relativity of border permeability is made equally clear by the ease with which economic elites cross borders in contrast to less educated and less prosperous people. This makes plain that borders function as imperfect filters reflecting both inclusive and exclusive policies, which differentiate between types of people, materials, and motives. The concept of border permeability is therefore not necessarily linked to a pursuit of freer movement. Rather it is a variable category that fluctuates between relative closure and relative openness.

To ensure profitable and ethical treatment of increasingly intense transborder flows, the goal must be to construct "good" borders. Toward this end, many would argue that "good" borders generally feature open communication, formal demarcation agreements, standing boundary commissions, accessible transportation links, and a minimal military or police presence. But "good" borders may also hinder harmful actions by neighbors or factions within neighboring states. Therefore a border's ability to impede transboundary flows may be as essential and constructive as its bridging function. The threat of terrorism and drug trafficking combined with global economic integration and labor migration makes this point eminently clear. As state power is expanded in the name of security and simultaneously reduced under pressures for greater global economic integration, the relationship between borders and sovereignty remains profoundly unstable.

Borders and new landscapes of sovereignty

Despite incessant government pronouncements to the contrary, the doctrine of sovereignty has never been absolute or equally applied. Indeed, scholars have taken major strides in thinking beyond the parameters of the nation-state structure and its prescribed ideals of territorial sovereignty. The core of these new perspectives revolves around global processes taking place at subnational levels, as well as beyond the limits of state territory. In doing so, these processes contribute to new landscapes of sovereignty that disrupt the notion of mutually exclusive domains for the state and global scales.

Traditional approaches to border studies often interpreted the lines that divide states and other geopolitical entities as either naturally or historically "given." Rather than stabilizing into rigid and legitimated spatial dividers, borders are essentially evolving practices and institutions subject to the influences of contingent events and ideas. Today, such influences abound and include new ideals of human rights (reproductive rights, sexual orientation equality, rights to water), neoliberal economic practices (free movement of labor, outsourcing, privatization of state enterprises), and new asymmetrical security threats (terrorist networks, trans-state insurgencies, criminal cartels). Although some scholars relegated nationalism to the dustbin of history in the early 1990s, various new supranational, revived-national, and ambitious subnational identities have emerged and are emerging today replete with claims to sovereignty and requirements for borders.

New landscapes of sovereignty are making visible the complex networks linking local, provincial, national, regional, and global systems. They offer venues for monitoring processes of contestation, integration, co-optation, cooperation, inclusion, and exclusion. These venues subvert conceptions of the state as fixed and bounded. Instead of seeing the state as a container with clear distinctions between "inside" and "outside," the local, the national, and the "inter-" or "trans-" state are seen as increasingly flexible and interlinked. These process-orientated approaches to bordering reveal the very idea of sovereignty as flexible and fluid, and once again highlight the multifunctionality of borders as barriers, bridges, and filters.

Contingent sovereignty

The notion of "contingent sovereignty" illustrates the evolving nature of the nation-state system. Among border scholars, contingent sovereignty refers to the idea that territorial sovereignty, traditionally equated with the inviolable authority of the state, is being challenged by numerous groups arguing that violations of human rights or the proliferation of weapons of mass destruction compel international action. For proponents of contingent sovereignty, states in serious violation of global norms forfeit their territorial sovereignty. The international community is then empowered to intervene in the domestic affairs of these states to uphold various conventions and treaties. Once a violator's territorial sovereignty has been breached and the transgression addressed, a period of monitoring and oversight ensues. The offending state may (re)earn its territorial sovereignty contingent upon "good behavior" as determined by some international organization. Examples include the Road Map for Peace in the Middle East, the Good Friday Agreement in Northern Ireland, the Naivasha Agreement for Sudan, the Baker Peace Accords for Bosnia, and UNSC Resolution 1272 for East Timor. These new conceptions of state sovereignty and calls for global enforcement of accepted norms are, nevertheless, a far cry from a patently new system of global sociospatial organization. Can one imagine China or the United States subjecting themselves to the process outlined above? Contingent sovereignty and its prescribed violability of international borders, therefore, remain applicable to only those states lacking the power to prevent it.

The problematic nature of contingent sovereignty was laid bare by the international response to the so-called Arab Spring revolts in

2011. Following a UN resolution, members of the North Atlantic Treaty Organization (NATO) launched air strikes supporting rebels seeking to overthrow the Libyan dictator Muammar Gaddafi. Although technically a violation of Libyan sovereignty since it was a recognized UN member state, NATO leaders justified the strikes on the grounds that Gaddafi loyalists were engaging in human rights abuses. Therefore the international community could intervene because it had a "responsibility to protect" civilians. The NATO strikes played a critical role in the rebellion's eventual success.

In contrast, the Syrian regime of Bashar al-Assad was also accused of brutally suppressing antigovernment protests, yet the international community consistently dismissed calls for intervention throughout 2011 and early 2012. Foreign intervention did, however, occur in Bahrain as the country's predominantly Sunni government violently cracked down on protestors from its predominantly Shiite population. Yet this intervention took the form of Saudi Arabia and other Sunni governments dispatching forces to assist the Bahraini regime against the protestors. This suggests the "responsibility to protect" doctrine may be applied when the offending regime is very weak but not if the offending regime has formidable military forces, such as North Korea, or is allied with global powers, such as Bahrain, which hosts a major American military installation. Indeed, even in cases of obvious and sustained human rights or international treaty violations, such as Sudan, Rwanda, or Myanmar (Burma), most states remain hesitant to risk conflict.

These instances of state-sponsored abuses against civilian populations fueled calls for new interpretations of sovereignty throughout the 1990s and 2000s. For many human rights activists, sovereignty rests with the people. As such, a state's sovereignty is contingent upon upholding the basic rights of its people. Intervention across international borders is justified if a state violates the human rights of its citizens. This view prioritizes the ideals of justice and human dignity over state sovereignty. A similar perspective is evident in emerging border constructs relating to minorities and indigenous peoples.

Minority territories and indigenous sovereignty

The nation-state system was largely built on the notion of *uti possedetis* (as you possess, so you may possess) and Western legal traditions of property and ownership (possession being nine-tenths of the law). The concept of "indigenous sovereignty" challenges these notions and exposes the racisms embedded in the historical, political, and legal treatment of minority and indigenous land claims. As a result, activists and scholars have called for reconfigurations of power both within and outside a law, right, and sovereignty paradigm. For many, premodern notions of overlapping polities, nonhierarchical power structures, and frontiers have no less standing than Western a priori ideals of territory, authority, and boundary. From this perspective, the injustices suffered by minority and indigenous populations during the process of state formation deserve redress.

For advocates of indigenous sovereignty, the unacknowledged negative impact of border formation on contemporary intrastate politics constitutes a moral quandary. Put simply, most states formed under highly undemocratic circumstances. The formation of state borders generally results from unequal power relationships that both reflect and cross various social boundaries. Even borders demarcated to facilitate the formation of democratic states and civic nations are rarely the product of democratic processes. Ironically, democratic politics are expected to emerge from democratic institutions tied to modern states despite the fact that there cannot be democracy until democratic institutions and state borders are established. Tension often remains over this original imposition of power by one group over another, which may generate lasting economic and sociopolitical inequalities. These inequalities have fueled numerous international conflicts as well as ethnic resentments and social injustices. In this sense, borders represent the "scars of history" not only physically in the landscape but also symbolically and metaphorically in the minds of various populations.

Many different forms of minority territory resulted from the varying cultural, economic, and political conditions that accompanied the process of state formation. Enclaves and exclaves are good examples. Enclaves are a portion of one state's territory completely surrounded by the territory of another state. An exclave, by contrast, is a territory belonging to one state that is not contiguous with the rest of the state. Most exclaves are also enclaves inside another state. Exceptions to these definitions would be exclaves that are not completely surrounded by another state (e.g., the Spanish exclave of Ceuta on the North African coast) or enclaves that do not belong to another state (e.g., San Marino inside Italy).

Enclaves and exclaves are quintessentially the "scars of history," often generating intense emotion among the populations of the adjacent states. The 2010 violence in southern Kyrgyzstan is a prime example of such simmering tensions. However, it is important to recognize that territorial structure is not the cause of the violence. Rather, violence and tension results from specific conditions and circumstances. Political alliances, governmental instability, economic hardship, and geopolitical leveraging by neighboring states provided the underlying causes for the Kyrgyz clashes with their Uzbek enclaves. That being said, one cannot deny the oddity of these enclave/exclave borders.

The Uzbek enclaves inside Kyrgyzstan, like many borders in the former Soviet Union, stem from the idiosyncratic efforts of Joseph Stalin to create a federal union from the multinational czarist empire. Fifteen larger ethnic groups, such as the Russians, Ukrainians, and Uzbeks, were given republics within the Soviet Union, while smaller ethnic groups, including the Karakalpaks,

Tartars, Abkhazians, Chechens, and dozens of others, received autonomous regions within another ethnic group's republic. Since the Soviet Union was highly centralized, these borders were largely symbolic and relatively unproblematic. This situation changed following the collapse of the Soviet Union as provincial borders became international borders and autonomous regions gained greater political significance. Bitter fighting between Armenians and Azerbaijanis over Nagorno-Karabakh, Georgia's disputes with South Ossetia and Abkhazia, and the recurring clashes between Chechens and Russians provide continuing evidence. Today similar tensions may be brewing in the Uighur and Tibetan autonomous regions in China.

Autonomous ethnic regions are not, however, unique to socialist states. The United States has a reservation system for much of its indigenous population. Some 310 of these areas exist in twenty-six different American states. These areas possess special jurisdictional rights. Minor crimes committed on the reservations are adjudicated by tribal councils or courts with limited sentencing capacity. Major crimes are investigated by federal law enforcement officials and adjudicated in federal courts. Gambling is the best-known result of special jurisdictional rights on reservations. Until the 1980s, gambling was legal only in Nevada, Atlantic City in New Jersey, and on river boats, but Native Americans have since seized upon their special tribal sovereignty to build casinos. Revenues from luring "tourists" to the reservations were hoped to counter the inordinately low standards of living common to these ethnic enclaves, though intense debate exists as to the distribution of gambling proceeds.

American Indian reservations are just one example of the complex relationship between national governments and indigenous peoples around the world. It is clear that state borders and national, ethnic, religious, and linguistic communities have rarely, if ever, directly aligned with one another. Yet in response to the new realities of global terrorism and migration, many states seek to harden their borders in an effort to reassert territorial sovereignty and create an environment favoring cultural purity and government-prescribed loyalty. Such efforts speak to the centrality of security in the twenty-first century, but due to globalization, security is increasingly linked to a range of transnational issues and processes. Recognizing this, some scholars have suggested that for states to be truly secure and profitable, they will have to adopt more internationalist and de-territorialized strategies.

Graduated and detached sovereignties

The growing significance of transnational corporations, organizations, and practices has compelled some states to create new substate borderings that facilitate neoliberal, global, economic networks. As a result, it appears that sovereignty over some segments of the economy, society, and territory are shifting away from the state to supranational, global, or private entities. This shift constitutes a dramatic restructuring of the relationship between borders, the governing, and the governed.

Such a restructuring is evident in the idea of "graduated sovereignty," which refers to a state's differential treatment of segments of its population or territory. Special economic zones (SEZs) are examples of such novel borderings where states have voluntarily limited their authority over specific spaces. SEZs (also called free trade zones, export processing zones, free economic zones) are substate districts subject to more liberal or relaxed economic regulations than the rest of the state in hopes of attracting foreign investment and spurring growth. They range from technical innovation zones to industrial parks to tourist districts. Despite these differences, each challenges notions of absolute territorial sovereignty and blurs the distinction between international and domestic borders. SEZs can be found in virtually every corner of the developing world. India, for example, has some 220 functioning SEZs with another 280 proposed by 2012.

The unique borderings resulting from these recalibrations of sovereignty are also present in developed regions. Provincial governments in the United States have tried to attract foreign businesses like BMW, Honda, or Toyota by offering significant tax reductions and other incentives. Research into these relationships between provincial governments and foreign companies has shown that many companies often pay considerably lower land, building, and equipment taxes than established local businesses.

The government of Malaysia has followed a similar approach to stimulate economic development. In addition to granting special privileges to its entrepreneurial class and foreign investors, the Malaysian government also limited the ability of laborers to unionize and allowed bonded labor in export-orientated industries. This created a multitiered system exerting stricter control over manual laborers but minimal regulation over businesses and investors. Such efforts to subordinate state sovereignty to economic interests suggest a growing incongruence between power and rigidly territorialized state authority.

Extraterritorial or detached expressions of sovereignty (control of places beyond the recognized limits of state jurisdiction) are additional examples of exceptional borderings. Detached sovereignties include things like transit portals in airports, embassies, and ships at sea. Although detached from their state, each functions to greater and lesser degrees as bounded sovereign territory of that state. Immigrant and refugee centers, as well as special detention facilities, often constitute extra-territorial sites of state authority and further challenge the spatial underpinnings of law enforcement and justice.

The U.S. Naval Base at Guantanamo Bay, Cuba (GITMO) is undoubtedly the best known of these extra-territorial sites. The expansion of the United States in the late nineteenth and early twentieth centuries created questions regarding the territorial scope of the U.S. Constitution. Initially limited to the

"United States proper," a 1901 Supreme Court ruling extended a limited number of constitutional rights to citizens living within overseas territories that were controlled by the United States but not intended to become states. During World War II, full constitutional rights were extended to American citizens on territory controlled by the United States but not part of the country proper (e.g., overseas military bases). Foreign nationals occupying such territory were, however, not covered.

This provided the basis for George W. Bush's administration's argument that GITMO was essentially a "rightless" zone for foreign nationals since it is foreign territory "leased" from Cuba. Rights to liberty, fair trial, and habeas corpus, as well as the Geneva Convention, were deemed inapplicable. Therefore, the base seemed ideally suited to serve as a detention center for suspected "enemy combatants" in the war on terror. The U.S. Supreme Court rejected this argument twice, and in 2009 President Barack Obama ordered the prison closed and detainees transferred to a prison in Illinois. By 2011, however, Congress countermanded this order by approving legislation, which prevented the transfer of GITMO prisoners to the mainland or other foreign countries. This effectively stopped the closure of the detention facility, leaving the fate of detainees as yet unsettled.

While controversy surrounds GITMO's status, the idea of extraterritorial jurisdiction can have positive connotations as well. These include the power of the International Criminal Court to prosecute people for war crimes, acts of genocide, and other crimes against humanity committed outside their home states. Yet the ability to prosecute alleged international criminals remains contingent upon the cooperation of sovereign states. As a result, accused criminals can find sanctuary in sympathetic states, such as Saudi Arabia offering sanctuary to dictator Zine el-Abidine Ben Ali of Tunisia. While the arrest of dictators such as Argentina's Augusto Pinochet and Liberia's Charles Taylor represent positive acts of extraterritorial justice, the related practice of "rendition" (the abduction and transfer of people to states practicing torture) poses serious ethical questions and reveals significant potential for abuse.

Emerging borderlands

Conventionally depicted on maps as solid black lines, borders convey the impression of finality and permanence. Yet significant portions of the approximately 308 land boundaries separating roughly 190 states remain undetermined. Vast new borderlands emerge if air and water are also considered. Fresh thinking is required as states and other actors negotiate sovereign control over these new realms. About 160 maritime borders are currently demarcated, but some 270 of these borders remain in question. The fact that only one-third of potential water borders have been firmly established reflects centuries of debate concerning the extent to which states may claim exclusive sovereignty over the seas.

Initially determined by the range of a seventeenth-century cannon, the territorial waters of states were set at three miles from shore. This rather loose standard held all waters beyond the three-mile coastal zone to be international waters subject to the *mare liberum* or "Freedom of the Seas" principle. Being outside the sovereign control of any state, the high seas remain among the last frontiers in the modern world. Nevertheless, in the early twentieth century, state interests in mineral resources, fish stocks, and pollution protection catalyzed various redefinitions of the limits of territorial waters. President Harry Truman's 1945 declaration of the continental shelf as the natural extension of national territory set a precedent for later seabed resource exploitation.

Other states have opted for a twelve-mile zone of maritime sovereignty, while still others claimed "exclusive economic zones" (EEZs) stretching as far as 200 nautical miles from the coastal low-water mark. Convened between 1967 and 1982, the United Nations Convention on the Law of the Sea (UNCLOS) introduced provisions for maritime sovereignty over EEZs and continental shelves, navigation and transit procedures, deep seabed mining, marine environment protection, exploration and research, and dispute resolution. Codified in 1982 and in effect since 1994, the UNCLOS has come under increasing pressure, especially as states seek to exploit oil and natural gas reserves beneath the seabed. An example can be found in the controversy surrounding offshore drilling in the Florida Straits between the United States and Cuba.

The United States currently bans drilling in its portion of the Florida Straits. By contrast, since 2002 Cuba has issued several leases within its portion of the Florida Straits and the Gulf of Mexico to foreign oil companies. Controversy swelled in 2008, when Vice President Dick Cheney claimed China was drilling in Cuban waters while U.S. waters remained closed. In actuality, China is not currently among the holders of Cuban offshore drilling leases. Oil and gas companies based in Spain, Norway, India, Malaysia, Venezuela, Vietnam, and Brazil hold these rights and are engaged in resource exploitation. In 2010, fear of losing out on valued resource deposits prompted President Obama to propose opening portions of the eastern seaboard and eastern Gulf Coast for new oil and gas exploration. Drilling would still be banned in the American portions of the Florida Straits, but competition for seabed resources is clearly a growing factor in the global balance of economic power.

The competing maritime claims in the South China Sea have threatened to become an even greater international issue. In addition to being one of the world's busiest shipping routes, these waters also support very productive fisheries. Additionally, oil and natural gas exploration has revealed the possibility of major undersea deposits and raised the commercial and strategic stakes considerably. As a result, there have been heated words and periodic naval skirmishes between rival claimant states, especially between China and its southern neighbors. The matter is further complicated by the presence of Indian and American naval, commercial, and scientific vessels.

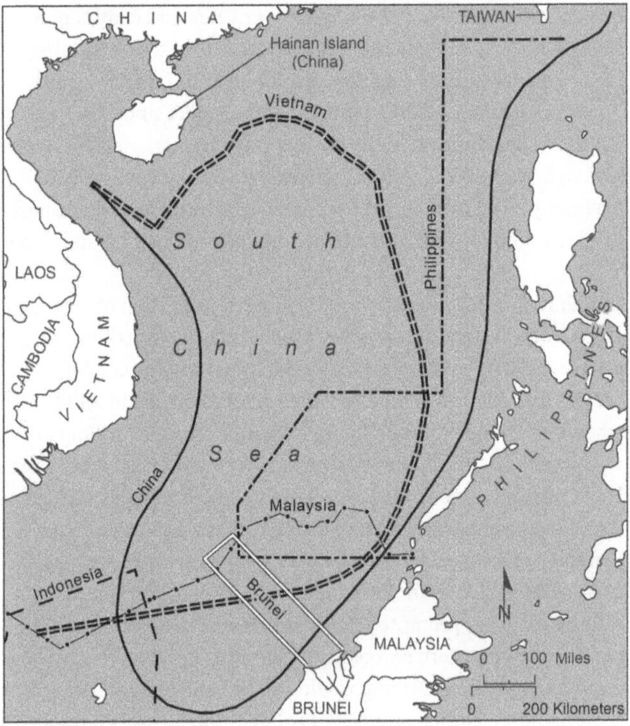

6. **Maritime claims in the South China Sea.**

Another maritime controversy has developed in the Arctic region as the steady retreat of ice pack opens the region for exploration and commerce and transforms its strategic importance. In addition to providing shorter shipping routes between many locations in the Northern Hemisphere, an increasingly ice-free Arctic would undoubtedly become a major commercial fishery, and initial exploration indicates significant undersea oil and gas deposits. This issue gained international prominence in 2007 when a Russian expedition used a submersible to plant a Russian flag on the seabed at the North Pole to claim the area and its resources as an extension of its sovereign territory. Other states bordering the

Arctic region soon launched expeditions to survey the seafloor and strengthen their naval capabilities in the region. Some have already filed official "continental shelf" claims based on the UNCLOS, but the issue of Arctic sovereignty remains unresolved.

The demarcation of airspace has also evolved with technological advances and raises profound questions concerning the viability of extending the ideals of territorial sovereignty skyward. Control over airspace generally corresponds with sovereignty over land and extends up to the Kármán Line, which marks the general upper limit of aerodynamic flight. Control of this space was irrelevant before the advent of air travel but became increasingly significant, both materially and symbolically, as national airlines became symbols of modernity and states vigorously protected their airspace rights. However, the deregulation of air travel and the promotion of the International Civil Aviation Organization (ICAO) since the 1960s have made states less assertive in protecting their national air-carriers and airspace. This supranational organization and cross-border coordination are vital, especially for airports in close proximity to borders. The ICAO, however, serves only as an arbitrator between states. States retain sovereignty over the air above their territories.

Airspace sovereignty therefore retains geopolitical significance. Governments have regularly used the enforcement and violation of airspace as a means of sending messages. Russian incursions into Lithuanian airspace, for example, signal Moscow's unwavering resolve to maintain its Kaliningrad exclave. France's refusal to allow American planes to cross its airspace en route to the 1986 bombing of Libya revealed to the world that Paris disapproved of the operation. Today, the geopolitical relevance of air-border security is enhanced by missile defense systems, radar monitoring, and satellite technology.

Euphemistically referred to as the final frontier, outer space gained clear geopolitical significance following the launch of the first satellite into orbit by the Soviet Union in 1957. This led to the Outer Space Treaty in 1967, which declared outer space, the moon, and other celestial objects to be "the province of all mankind" not subject to claims of state sovereignty. The treaty prohibits states from placing nuclear and other weapons of mass destruction anywhere in space. The construction of military bases or other military activities on the moon or other celestial bodies is also specifically forbidden. Notably, the treaty does not prohibit the deployment or use of conventional weapons in orbit around the Earth or moon, or just floating in space. This paved the way for the U.S. Strategic Defense Initiative, or "Star Wars" program, in the 1980s that caused great concern that de facto American sovereignty over outer space was imminent. Star Wars was never realized, but the United States, Russia, and China all currently possess Earth-based weapons systems capable of destroying an opponent's satellites should the need arise.

Although space itself is not subject to claims of state sovereignty, the Outer Space Treaty recognizes satellites and other spacecraft as carrying the sanctity of state sovereignty similar to ships at sea or embassies in foreign countries. Advances in satellite and remote sensing technologies have had a variety of cross-border commercial and cultural applications ranging from rapid dissemination of news via global media conglomerates to the availability of clear aerial images of much of the world's surface via the Internet. Satellites have obvious strategic applications for cross-border surveillance. Although overhead flights below the Kármán Line would violate state airspace, satellites orbit above the line; so detailed aerial images can be obtained without violating sovereignty. Although satellites themselves have not been weaponized, as far as we know, they still have direct military implications. The U.S. Global Positioning System (GPS) relies on a network of satellites in geosynchronous orbit to provide location information for planes and ships as well as to an increasing number of drivers and hikers. GPS also provides a significant advantage for the American military by providing extremely accurate targeting information for cruise missiles and other guided munitions. The European Union, Russia, China, and India are currently working to develop or improve their own GPS capabilities. These examples from air, sea, and space highlight how approaches to territory, sovereignty, and borders continue to evolve in response to new technologies, commercial possibilities, and geopolitical concerns.

Chapter 5
Border crossers and border crossings

People are crossing borders with unprecedented frequency and volume. Yet the processes and experiences of border-crossing are quite varied. Some borders are marked only by simple signage or entail just a brief identification check, while others require extensive documentation and searches. Variations in crossings create distinct borderlands or zones of transition and an encounter between two or more territories. Some borderlands provide settings where diversity is enjoyed and celebrated, while others are troublesome zones of tension and antagonism.

Migrants and refugees

Many contemporary social theorists suggest that the near pervasive processes of border-crossing are remaking social categories of belonging, especially the idea of citizenship. This discussion contrasts the negative connotations commonly associated with words such as immigrants, deportees, and refugees with more positive words such as cosmopolitans, jetsetters, and global citizens. The latter categories suggest the emergence of new spaces of interaction and connectivity that will enable transcendence of rigid sociospatial systems and give rise to free, self-managing, and self-enterprising individuals. Such a perspective textures our understanding of generally unquestioned ideals of citizenship and loyalty by demonstrating the mutability of categories such as insider or outsider and processes of inclusion or exclusion. Nevertheless, dichotomies of this nature remain and are increasingly prevalent in the world as evidenced in viewing the individual human body as a border site.

The idea of diplomatic immunity, for example, derives from the nation-state system but functions as a form of mobile sovereignty, which largely exempts diplomats from the legal jurisdiction of the state in which they physically reside. These individuals literally and figuratively transplant the sovereignty of their home country into the jurisdiction of the host state. They are "in" another country but not "of" that country and, as such, are largely excluded from the responsibilities and restrictions of the host society. Tourist, work, and education visas offer similar manifestations of achieving mobile citizenship and national identity by framing the body as sovereign space.

In contrast to the diplomat's immunity or the visa carrying tourist's acknowledged presence within a given territory, the unauthorized immigrant's body constitutes a violation of sovereignty and the very site of border enforcement through detention or expulsion. The difficulties facing unauthorized immigrants stand in stark contrast to the relative ease with which other people cross boundaries. Those with specialized skills, for example, often find borders very porous, while unskilled laborers generally confront greater resistance to their movement. For those concerned with migrant rights, borders constitute troublesome sites of vulnerability. Illegal immigrants obviously face the most extreme risks, but all border crossings provide potential venues for extortion, exploitation, and abuse. For some scholars and human rights activists, rigorous enforcement of stringent immigration policies violates human dignity and is therefore unethical. Given the costs of welfare, health care, law enforcement, education, and other social services, most governments would counter that limiting access to state benefits is necessary for a functioning society.

Russia is a prime example of such a migration debate. Since the collapse of the Soviet Union, Russia has emerged as a "migration magnet." Legal and undocumented immigrants flocked to Russia from countries stretching from Africa to East Asia. Even more immigrants come from within the Commonwealth of Independent States. Their legal status is uncertain due to agreements allowing visa-free travel between former Soviet republics. Consensus estimates for the number of illegal migrants in Russia currently range between 3.0 to 3.5 million, with plausible estimates as high as six million. Some migrants are simply using Russia as a transit point on their way to the European Union, but others consider Russia a "land of opportunity." Despite this, immigrant populations face significant obstacles as upwards of 60 percent of Russians support such slogans as "Russia is for [ethnic] Russians." Refugee status is rare, many migrants are denied the right to work or access to social services, and the rate of ethnically motivated violence against migrants is rising. Yet, demographic projections paint a dire picture that could temper policies restricting immigration. The combination of a "brain drain" of some of Russia's best and brightest to western countries and the "graying" of Russia has led some (mostly social liberals) to advocate "replacement migration"; that is, immigration to offset the decline and aging of Russia's population. So far, immigration opponents are generally winning the debate by vociferously arguing that "foreigners" contribute inordinately to criminality and erode the purity of Russian culture.

Immigration restrictions have also generated intense controversy in Canada, the United States, Australia, Japan, and across Europe. Recent efforts to restrict African emigration into the European Union, for example, have involved the construction of new fortifications along Spain's southern borders as well as increased patrols in the waters separating Europe from Africa. These concrete actions are coupled with a symbolic contrasting of Spanish/European identities versus Moroccan/African identities. These efforts reveal how borders are far more than lines on the map or locations in a landscape but rather constitute modifiable processes of exclusion, alienation, and distinction. The categorical definition of the "welcome" versus the "unwelcome" is often contingent upon specific factors relating to the bordered reality of states.

The very idea of refugees provides one example. The United Nations defines refugees as those crossing borders due to a "well-founded fear of being persecuted for reasons of race, religion, nationality, membership of a particular social group, or political opinion." Refugee status is therefore explicitly conditional upon being outside the country of one's citizenship and unable or unwilling to rely on protection from that state's legal system. As a result, refugees are entitled to asylum, or the right to remain in a foreign state without the possibility of extradition to their home state. This highlights the harsh reality of differential treatment for those crossing state borders. While refugee status is available to those escaping the oppression of communist or religious dictators, economic migrants fleeing poverty or environmental degradation are not afforded this right.

7. UN peacekeepers from Thailand greet children at a refugee camp in Darfur, Sudan, 2011.

Such social and legal distinctions between citizen and noncitizen, immigrant and refugee, and voluntary and involuntary migrant constitute geographic categories of belonging, between those "in place" and those "out of place." Borders play central roles in institutionalizing these differences and their corresponding geographic expressions. Yet, identities are never completely contained by borders.

Transborder identities and communities

Migration studies have traditionally emphasized the monodirectional movements of people to permanent settlements in new places, such as the migration of Europeans to the Americas. Today, however, patterns of cyclical, return, or seasonal migration between states have become increasingly common. Such migrations inherently involve multiple border-crossings over the course of a lifetime or even during a given year. Coupled with advances in communication and transportation technologies, these cyclical patterns of migration enable the formation of dense social networks between spatially dispersed groups to a far greater degree than previously. These new migration patterns entail profound economic, cultural, political, and social changes and blur the distinction between "sending" and "receiving" communities.

These "transnational social fields" can be organized to facilitate collective action beyond the traditional nation-state system. The idea of hybridized identity, wherein one's ethnic heritage is attached to a base of current state affiliation (Irish American, Korean Kazakhstani, etc.), is today quite commonplace. In most instances, such constructions are banal and serve only to bolster a sense of individuality. In other instances, the concept of sovereignty is profoundly tested as diasporas (usually defined as ethnic groups living beyond the borders of their national homelands with desires to someday return) are increasingly able to influence the domestic debates in their homelands, and homeland-elites are better able to mobilize diasporas. The apparent durability of current transnational social fields draws the very definition of diaspora into question, as many dispersed groups display minimal interest in returning to their historic homelands. They choose instead a hybridized identity that constitutes a status of national belonging to "both/and" rather than "either/or."

Examples of such transnational social fields include high-profile groups such as the "Jewish Lobby" in the United States, Algerian populations in France, Turkish "guest workers" in Germany, and the Armenian diasporic communities in many states. Less publicized groups such as Tajik labor migrants in Russia, Lebanese merchant groups in Africa, and growing Asian communities in Australia, among many others, offer case studies in radically different socioeconomic, political, and cultural circumstances. In each case, political influence, cultural infusions, and economic remittances have affected both the sending and receiving societies. These links have created dependencies and vested power in new places and peoples.

The growing influence of transnational social fields has been especially evident across Central Eurasia since the collapse of the Soviet Union. For example, diasporic networks became particularly significant for Armenian elites, both domestically and abroad. Some Armenian émigrés were highly nationalized and wished to return migrate to the newly independent Armenian state formed when the Soviet Union dissolved. Others in this diasporic community were settled in their new locations abroad. As a result, diasporic organizations, like the Armenian Pan-National Movement, were established to manage the flow of remittance monies, form development programs, organize educational exchanges, and coordinate cultural transfers. Seeing the success of these groups, other newly independent nations of Central Eurasia established their own diasporic outreach associations, each harnessing some level of national attachment among their dispersed communities. Examples include the World Azerbaijani Congress, World Association of Kazakhs, and the Crimean Tatar National Assembly to name but a few.

As in other developing regions, such transborder links raise important questions concerning state tolerance of transnationalism, especially when "return migration" is not the goal. In these cases, a "stretching of the homeland" is envisioned to offer opportunities for dispersed peoples to own property, sponsor programs, attain education, and avail themselves of the culture of their historic homeland without actually migrating or assuming the responsibilities of citizenship within that state. Such a process is extremely complex in the post–September 11 world, as Muslims confront discrimination in western states and geopolitical relations have become strained between governments with divergent positions on the "war on terror." Given the varied roles of dispersed peoples in the contemporary international system, diasporas likely have greater political weight than ever before.

The historical legacy of the British Empire in South Asia continues through transnational social fields occasioned by the steady flow of immigrants from India and Pakistan to the United Kingdom. On the positive side of the ledger, these migration streams provide a valuable labor source with the language skills and cultural familiarity to contribute to British society, as well as adding another layer to an already cosmopolitan cultural landscape. On the negative side of the ledger, however, postcolonial antagonisms and geopolitical tensions between India and Pakistan have been reproduced among their respective émigré communities. In this case, postcolonial migration helped spread existing racisms/jingoisms and transplant the tensions of the Indian-Pakistani border to British cities. This highlights the importance of understanding how new urban or substate borders obtain their significance from the identities carried within individuals and groups. It also raises important questions concerning the value of border permeability. Indeed, the crossing of borders has the potential to bring great benefits to states, provinces, and municipalities, but such crossings also have the potential to reproduce various forms of socioeconomic injustices and even violence.

Insurgents and terrorists

War has traditionally been understood as a military conflict between two sovereign states, like the Iran-Iraq War or the Falklands War between the United Kingdom and Argentina during the 1980s. Yet, most contemporary wars have involved larger numbers of states and various nonstate combatants. Some wars involve larger alliances or coalitions of states taking action against regimes deemed to have violated international law or norms. These types of efforts are generally approved by the international community. NATO-led operations in Afghanistan and Libya are two recent examples, as is the UN-authorized coalition in the Gulf War in 1990–91. Other conflicts involve groups of states initiating military operations without international sanction, such as the U.S.-led invasion of Iraq in 2003 and the repeated foreign incursions into the Democratic Republic of Congo.

Despite the dramatic nature of these large-scale military operations, contemporary conflict is more likely to involve smaller-scale border crossings by nonstate combatants. Quite often, these are part of conflicts that appear confined within a single state. Terms such as civil war, internal conflict, and domestic strife generally refer to violence between competing factions, usually a government and an insurgent group, within the borders of a state. The domestic nature of these conflicts is misleading since the majority of insurgent groups active since 1945 have operated beyond the borders of their home state. Assassinations, bombings, supply bases, training camps, fund-raising networks, and other operations usually extend into neighboring or distant states.

The Kurdish Workers Party (PKK) illustrates the cross-border nature of most insurgent movements. The PKK has engaged in a long-term effort to carve a Kurdish state from the territory of contemporary Turkey. These efforts have involved attacks in Turkey and against Turkish interests abroad. The PKK draws support from the Kurdish diaspora community and has trained its combatants in camps in northern Iraq and Iran. The U.S.-imposed no-fly-zone over northern Iraq inadvertently provided sanctuary for PKK bases and developed into a major source of tension. In response, Turkey refused to allow the United States to use Turkish territory during the 2003 invasion of Iraq. Turkey has subsequently violated Iraqi airspace on several occasions to interdict the PKK threat and even invaded Iraq for eight days in February 2008. This example demonstrates the importance of understanding contemporary "civil" conflicts within a broader international context. Modern insurgent, revolutionary, and secessionist movements are rarely fully domestic projects. Most directly involve some type of cross-border activity, thereby blurring the line between civil and international conflict.

Similarly, global terrorist networks undermine basic premises of the modern state system by relying on trans-state funding and communication to circumvent state security apparatuses. Terrorists are generally prosecuted in state legal systems like common criminals, though the scope of criminality extends far beyond the borders of any particular state. Terrorists who kill innocent people, for example, will be charged with murder in the state where the attack occurs. Yet the preparations for that crime likely extend abroad. The individual may be funded and instructed/inspired to carry out the act by an organization operating in a different country or perhaps in the murky spaces of the Internet. The capacity to prevent future acts of this nature will undoubtedly require action beyond the borders of the state in which the attack occurred. Ironically, states often further undermine international norms by launching preemptory or retaliatory attacks that transgress sovereign state borders, such as U.S. attacks against targets in Libya, Sudan, Afghanistan, Bosnia, and Serbia in the 1980s and 1990s.

The Al Qaeda organization is clearly the most notorious of contemporary terror groups. Although its actual membership is quite small, the Al Qaeda "brand" is currently embraced by various religious groups around the world to generate credibility, visibility, and funding for their causes. Instead of defending some homeland and patrolling borders, these terrorist groups are cellular entities detached from the geopolitical norms of the nation-state system. The de-territorialized threats of these "nonstate actors" compelled President George W. Bush to declare war on the very tactic of terror. Such a war is unique in world history as the tactic cannot surrender.

The conflict along the Afghanistan-Pakistan borderlands demonstrates the complexity of the war on terror through its combative and cooperative intermingling of state militaries, insurgent militias, international coalitions, and terrorist movements. An American-led NATO force has been battling Al Qaeda and its Taliban allies in Afghanistan since 2001, but many Al Qaeda/Taliban forces escaped by simply crossing the border into northwestern Pakistan. Although part of Pakistan's sovereign territory, local tribes in these regions enjoyed considerable autonomy. Sharing many ethnic and cultural ties, these tribal leaders were generally sympathetic toward the Taliban. These areas soon evolved into a sanctuary for Al Qaeda/Taliban forces to launch guerilla attacks back in Afghanistan or prepare terrorist attacks overseas. American forces were officially forbidden from crossing the border but nonetheless launched an increasing number of missile attacks from remotely piloted drone aircraft patrolling, and officially violating Pakistani airspace. These cross-border operations climaxed with the helicopter incursion of American commandos deep into Pakistan that resulted in the death of Al Qaeda leader Osama bin Laden in 2011. Human rights activists, international lawyers, and national security experts debated the morality and legality of the bin Laden operation and the drone strikes. This highlights the subjective nature of determining whether groups or individuals should be classified as insurgents versus criminals, and speaks to the complexity of transborder crime and the myriad issues involved in efforts to interdict criminal activity that spans state territory. Regardless, the United States appears to have expanded its drone program in response to civil strife in Yemen, Somalia, Libya, and perhaps other countries.

Criminals and police

Most states possess legal systems in which jurisdictional authority is based on territory. The borders between various national, provincial, and local courts help ensure the consistent and efficient enforcement of laws, punishment of criminals, and compensation for victims. Nevertheless, the idea of territorial sovereignty and the borders that demarcate jurisdiction often hamper efforts to stop criminal activity both internationally and domestically. Despite efforts to create and enforce some basic norms, legal systems and law enforcement around the world remain fragmented, diverse, and in some areas largely absent.

In some countries, legal systems are highly centralized, so jurisdictions have little practical effect. Other countries have more decentralized law enforcement comprising interconnected parts. These usually include a foundation of legal thought to define criminal behavior, a police system to enforce the laws deriving from that foundation, a court system to apply the law, and a corrections system to either punish or reform the convicted. In the United States, for example, the fifty states have the right to craft distinct codes of law and maintain separate police forces in accordance with the Constitution and Supreme Court decisions.

Despite efforts to implement specific universal principles of human rights and international behavior, the international system lacks a police force or a supreme authority to judge violations and determine punishments. Instead, it relies on state law enforcement agencies to cooperate and share information in their efforts to combat transborder criminals. INTERPOL and EUROPOL are coordinating bodies with limited authorities. Their duties traditionally focus on facilitating extradition and investigatory cooperation.

Although not without controversy, the past two decades have seen individual states formulate a variety of new legal procedures to address the cross-border realities of crime. Extradition and investigation were once the prime drivers of regular cooperation in the realm of international criminal justice, but increasingly new policies, courts, and detention facilities are emerging in response to new, expanded threats. The United States, for example, has moved toward unprecedented intelligence sharing between the FBI (domestic law enforcement), the CIA (foreign intelligence operations), and foreign governments. Each act of information exchange blurs the former legal distinctions between domestic and international security operations. The Department of Homeland Security is itself a border-related agency whose purpose is to ensure the barrier/filter functions of U.S. boundaries and facilitate the cooperation and implementation of best practices across the internal borders of the country.

Despite these efforts, cross-border crime remains broad in scope and difficult to interdict. Various forms of trafficking and smuggling, for example, are globally pervasive cross-border activities that utilize every conceivable mode of transport from cars, boats, and airplanes to pedestrian tunnels, paths, and river fords. The diversity of things to be smuggled and the means to smuggle them create a near impossible task for law enforcement. Although easy to detect, ships at sea are regarded as sovereign territory of the commissioning state. Acting on intelligence from the United States, the Spanish navy intercepted an unflagged North Korean freighter shipping Scud missiles to Yemen in 2002. At the time, North Korea was designated as a state sponsor of terrorism and, although a U.S. ally, Yemen was rife with political and ethnic strife. Although disapproving of the weapons transfer, the United States reluctantly allowed the delivery since according to international law, Yemen was a sovereign state legally entitled to purchase conventional weapons, and the freighter was considered sovereign property of North Korea.

While arms trafficking generally requires larger means of transport, drugs do not. Smaller searches of individual humans have proven profoundly challenging for law enforcement officials, as smugglers have employed "mules," or people who ingest or otherwise conceal contraband within their own bodies to evade detection at the border. Drug trafficking is especially troublesome for governments throughout the Americas. Fueled by demand for illegal narcotics in North America, drug cartels pose serious threats to border control agencies and even the territorial sovereignty of some countries in Central and South America. Beginning in the 1970s, Colombian drug cartels became so powerful from growing and smuggling cocaine that they established effective territorial control over portions of the country. After significant violence, Colombian authorities were largely successful in breaking up these cartels during the 1990s. Unfortunately, control over the drug trade merely shifted to other groups. The Revolutionary Armed Forces of Colombia (FARC) insurgent group filled the void left by the cartels and established de facto control over a sizeable territory in its fight against the Colombian government.

More recently, drug cartels in Mexico have dominated the narcotics trade, resulting in widespread violence between competing cartels and government forces. It appears that the Mexican government has lost effective control of portions of the country as this narco-violence resulted in the death of an estimated 11,000 people during 2010. This number is three to four times higher than the number of civilian and military deaths reported in Iraq during the same year. This tragic situation highlights the connection between drug and weapons trafficking as both FARC and the Mexican cartels rely on profits from the drug trade to purchase illegal weapons, fighters, and other supplies. Drug profits also sustain smuggling operations by funding the bribery of government officials and border authorities. This nexus of criminal drug gangs and weapons smuggling that fuels rebel groups has the effect of blurring the distinction between civilian criminal enterprises and violent political insurgencies. Unfortunately, demand for narcotics across Europe and Asia has also helped fuel insurgencies in Central and Southeast Asia.

The related enterprises of human smuggling and trafficking enable illegal flows of people across borders. Human smuggling involves assisting a person to gain unauthorized entry into a foreign country, usually in exchange for payment. This could involve serving as a guide to escort migrants across the border clandestinely or providing fraudulent documents to pass through an official entry point. Although taking different forms ranging from temporary bonded servitude to forced labor, human trafficking refers to the actual trade in human beings as property and is often termed "modern-day slavery." The people bought and sold are generally younger adults/teenagers or even younger children who are exploited for work ranging from domestic servants, sweatshop laborers, soldiers, or sex workers. Although the process of trafficking does not require the crossing of an international border, in practice most human trafficking operations have a trans-state dimension.

Other illicit activities, such as money laundering in "off-shore" banks, are complex financial crimes that require the crossing of borders. The infringement of intellectual property and patent rights is even more complex and seems to occur throughout the world with near impunity. Those "pirating" movies, music, and other forms of intellectual property should technically be arrested and prosecuted by state authorities. However, international borders greatly reduce this possibility. Chinese markets, for example, are filled with goods bearing the logos of prominent companies. Most of these sales occur without the permission of the brand-holding corporation, thereby making the sale of such items a crime.

Another example of intellectual property-rights infringement may be found in copycat pharmaceuticals distributed in developing countries. Although resulting in cheaper medicines that are more accessible for poorer people, patent-holding companies will have diminished global market share and profits to fund future research and development. As a consequence, the development of new treatments and medicines for a variety of diseases may be slowed or stopped entirely. Even if a reliable venue existed to address these infringements, the prosecution of companies replicating patented drugs without approval raises the ethical quandary of restricting access to medicines in impoverished regions.

Fortunately, most cross-border crimes are not so ambiguous. Computer hackers, for example, prowl the virtually unbounded realm of cyberspace in an effort to wreak havoc on public and private computer systems. These activities may appear as simply the mischief of young pranksters, but they have the potential to disrupt global processes and cause catastrophic damage. This problem is so profound that legislation has been advanced in the United States to provide the president with an Internet "kill switch" should a cataclysmic cyber-attack broadly threaten energy, transportation, or other vital systems. This kill switch represents a cyber-border around the United States not unlike that used more aggressively to regulate online content within China and many other non- or quasi-democratic regimes.

In a related vein, the Stop Online Piracy Act (SOPA) proposed to the U.S. Congress in January 2012 catalyzed a major controversy that pitted groups and companies like the Motion Picture Association of America, the Recording Industry Association of America, Nike, and Viacom, with interests in protecting intellectual property rights, against Google, Wikipedia, YouTube, Facebook, and other interests espousing a desire to protect free speech and innovation. U.S. politicians broke with normal party affiliations on the issue as it raised questions of governmental capacity to censor online content and bypass liability protections afforded to Internet sites by the Digital Millennium Copyright Act. To make their case known, Wikipedia and an estimated

7,000 other Internet sites shut down for a 24-hour period and Google blotted out its home page logo. The bill was postponed and its ultimate fate remains uncertain, but the issue of online piracy continues to be a prevalent concern.

Pirates of quite another sort constitute a distinct category of criminal actors operating around the margins of sovereign territory. These criminals present a direct threat to international shipping in certain parts of the world. Seizures of oil tankers and other cargo ships off the coast of Somalia have become common in recent years. This threatens to undermine the freedom of maritime transportation and again highlights the lack of sovereign authority over the seas.

Beyond their threat to high-seas commerce, Somalia's pirates also impact the function of their own state. Somalia's recent history spans famine and social division among warlords during the

8. U.S. military personnel approach suspected Somali pirates in the Gulf of Aden, 2009.

late 1980s, foreign intervention in the early 1990s, indigenous insurgencies bolstered by global ideologies of radical Islam during the 2000s, Ethiopia's invasion in 2006, and finally the appearance of postmodern piracy on the high seas. Somalia retains its status as a sovereign state, but the government is basically impotent. Indeed, many categorize Somalia as a "failed state" whose territory is actually controlled by various insurgent movements, clan militias, or pirate groups. The notion of a failed state has practical and moral implications for the role of borders as containers of sovereignty, when the government is incapable of exercising that sovereignty. Coupled with drought, grinding poverty, and food shortages, Somalia's borders seem more like a prison than expressions of state sovereignty.

Tourists, tourism, and borders

While the negative aspects of border crossings generate headlines, the majority of border crossings are generally positive in nature. Tourism is one of the most prominent forms of border crossing with immense economic impact, but like all sociocultural processes, there are two sides to the coin. Tourism involves travel motivated by the pursuit of fun, relaxation, and new experiences. Obviously, tourism can bring great benefit to travel destinations, but the negative side effects and corresponding ethical questions are too often overlooked.

Tourism is commonly defined in terms of distance traveled and time spent away from home, and invariably involves crossing one or more borders. These borders could be municipal, county, provincial, or international, but the crossing of any border speaks to the experiential side of tourism and the idea of leaving your daily routine and territory. For many the excitement of the trip is enhanced by the crossing of a highly visible political or cultural border. Tourism studies have shown that border crossing actually adds to the perceived distance of a trip—the more draconian the process of traversing the border the greater the perceived distance. Travel within democratic, developed states entails few complications, but the required travel documentation within less developed or nondemocratic states may be as extensive as international trips.

Crossing international borders subjects travelers to the scrutiny of both their country of origin and their destination. This scrutiny is symbolized by the documents required to prove individual identity, country of citizenship, and by extension one's rightful place in the world. Passports, visas, and other identification documents have become increasingly important since the September 11 terrorist attacks. Originally travelers carried simple letters of introduction, but the emergence of nation-states, greater emphasis on experts' claims to objective knowledge, and increasing bureaucratic centralization led to new rules governing international movement. Gradually, the passport gained acceptance as the most trustworthy proof of identity, even more reliable than the testimony of its holder. Modern passports are increasingly loaded with personal travel and citizenship data, anticounterfeiting technologies, and biometric tracking functions. Clearly, requirements for identity documents to legally cross borders are more extensive today that at any point in history.

Once satisfying these bureaucratic requirements, tourists often seek out markers of territorial difference, for example, by posing next to signs, fences, walls, or other types of political boundary markers. Some borders have actually become tourist destinations in and of themselves. Niagara Falls (U.S./Canada), Victoria Falls (Zambia/Zimbabwe) and Iguazu Falls (Brazil/Argentina) combine stunning natural attractions and international borders. Relict borders such as the Berlin Wall, Hadrian's Wall, and the Great Wall of China attract thousands of people annually. Even the odd meeting points of multiple political territories have become popular tourist destinations, such as the Four Corners Monument where Utah, Colorado, New Mexico, and Arizona converge, the Tri-point Monument where Finland, Sweden, and Norway meet, and the Triple Frontier of Argentina, Brazil, and Paraguay.

Through the process of border crossing, tourists traverse different sovereign legal regimes. This may result in travelers gaining or losing certain freedoms or restrictions. Some of these might involve drastic differences regarding basic personal freedoms (mobility, speech, religion, sexuality), while others involve the accessibility of certain goods and services (alcohol, pharmaceuticals, gambling, prostitution). Tourist destinations often market these different legal regimes to attract specific tourist populations. Tourists may also exploit different currency zones to enhance their purchasing power, especially in borderland regions. Catering to tourist desires and expectations, in turn, has required adaptations that often challenge the authenticity of local cultures and catalyze divergence from national ideals as borderland economies become increasingly dependent on cross-border flows. For many observers, the growth of tourism suggests an increasingly trans-state future.

Chapter 6
Cross-border institutions and systems

Cross-border institutions and systems have become far more pervasive and central to the lives of more people today than in any previous era. The existence of these cross-border networks undermines the common assumption that state borders and societal borders are, or at least should be, congruent. Environmental concerns, public health issues, and flows of information, for example, are only partially subject to the power of borders. Yet from another perspective, borders clearly impact the ecological, medical, and intellectual realties of human existence. These cross-border institutions and systems are rather ironically inherently resistant to and profoundly influenced by the bounding effects of political borders. Indeed, the complexity of border studies is perhaps no more apparent than in the realm of cross-border institutions and systems.

Ideas and information

Thanks to new communication and transportation technologies, ideas and information cross borders more readily today than at any point in history. Indeed, the emergence of cyberspace and the global reach of western media broadcasts challenge state sovereignty because of the inability of territorial borders to impede these flows and exchanges of information and ideas. But despite changing technologies, this reality is not entirely new.

The British Broadcasting Corporation's World Service and the U.S. government's Radio Free Europe and Voice of America, for example, transmitted western ideals to populations in communist countries during the Cold War. Even with these early examples, it is clear that cross-border exchanges of ideas and information occur with unprecedented speed, volume, and ease in the contemporary world. This has profound and contradictory implications for both political and social borders.

For example, modern telecommunication satellites, though lacking weapons, represent a type of strategic asset through their capacity to transmit information and influence public opinion. Western companies, like BBC, News Corporation, CNN, Deutsche Welle, and France24, hold large shares of the satellite television broadcasting market and are therefore in key positions to shape cultural mores and political ideals. Broadcasting the "news" from a particular ideological perspective speaks to the power to shape hearts and minds. In combination with film, television, and music, the global media network is largely controlled by western corporations and wields unprecedented transborder influence. Regional outlets have sought to counter this western dominance with their own broadcasting networks, such as Al Jazeera and Al Arabiya across Arab-speaking countries. Yet, even these flows are subject to bordering as numerous states censor or block satellite communications.

New "cyberspaces" of human interaction are also forming as the Internet enables novel exchanges of ideas and information. The seemingly unbounded nature of cyberspace offers a vast array of new opportunities for linkages among those with Internet access. Internet communities provide evidence for a detachment of identity and belonging from the bounded limits of state territory and sovereignty. The premise of "shared interests" has given rise to a wide range of new international communities. Some of these online communities are global humanitarian, environmental, or religious movements with specific ideological objectives, such as the International Committee of the Red Cross, Greenpeace, or

9. Nomads in Mongolia can stay connected and informed via communications satellites.

nebulous Islamist extremist groups. Others are simply intended for socializing, entertainment, and leisure, such as online dating services, gaming groups, or social network websites. In most cases, these new links encourage constructive and healthy cross-border dialogues. Unfortunately, like all identities, even these cyber communities can develop privileged and hierarchical structures with potentials for exclusion, bigotry, and perhaps human rights abuses. The generally open nature of cyberspace serves the harmful cross-border causes of Al Qaeda and computer hackers, just as it serves the humanitarian aims of Amnesty International and Reporters Without Borders.

The ability of groups to transcend space and traditional border barriers poses a challenge for governments around the world. Some even regard the Internet as the primary agent in flattening the world and erasing state borders. Yet despite its apparent openness, there is a clear geography of Internet censorship around the globe. Many of the nearly two billion Internet users face government restrictions when downloading or posting information online as states invest considerable resources toward monitoring and controlling its content and activity. Governments block websites, monitor chat rooms, and harass users of cyber cafes, among other limitations on cyber freedom. Democratic states debate the legality of implementing Internet restrictions raising difficult ethical issues concerning the proper balance between freedom of speech and preventing the distribution of reprehensible content, such as child pornography. Unconcerned with the dictates of free speech, nondemocratic governments have implemented strict censorship programs to block political dissent and alternate viewpoints. Given the massive and decentralized nature of Internet traffic, the effectiveness of such programs is unclear but certain governments clearly intend to regulate online flows of ideas and information.

The so-called Great Firewall of China offers a prime example. Beginning in 2003, China's communist-controlled government launched Operation Golden Shield to block access to certain Internet sites, programs, and content. The project also tries to censor websites considered politically subversive by Communist authorities. These could include websites advocating democratic reforms in China, independence for Tibet, or human rights improvements. China also blocks access to YouTube, Twitter, and several other social networking sites, fearing they could be used to document human rights abuses or organize antigovernment protests.

While China's censorship has been generally successful, this is not the case for all governments. In January 2011, the reign of Tunisian dictator Zine el-Abidine Ben Ali abruptly ended after twenty-three years. The uprising that pushed this dictator from power was not, however, driven by an ideologically motivated insurgent group but rather by general discontent and anger toward the government spread through the social networking website Facebook. After a police officer confiscated his produce cart, a young Tunisian named

Mohamed Bouazizi protested by setting himself ablaze in front of the regional governor's office. Photos and videos of subsequent protests were uploaded from cell phones to Facebook and succeeded in stirring the population to broader, sustained action. In response, government authorities implemented a program to steal online passwords and login information thereby enabling them to block further use of the social networking site, as well as potentially identifying those involved in the movement. Protestors were undeterred and created new accounts to organize their revolt. Eventually, the government succumbed to public pressure, and Ben Ali fled the country. Military strategists, government leaders, and political dissidents have long recognized the importance of information and propaganda, but this may be the first successful "cyber revolution."

While the power of cyberspace is manifest through the role of Facebook and other Internet sites in the Arab Spring revolts of 2011, state borders continue to shape and reflect a variety of social ideas and norms. Any consideration of borders must acknowledge the lines that divide the everyday social practices of males and females. The most obvious of these are gender-specific spaces designed to separate the sexes, often reflecting different gender roles derived from religious customs. Many of these attract little attention, like the creation of separate spaces to differentiate how and where women and men may worship. In other settings, gendered spaces have generated greater controversy. The Islamic custom of women wearing veils or other coverings, for example, has sparked significant debate in many western countries with Muslim minorities. For some, the veil simply reflects the ideals of modesty, fidelity, and marriage. Others interpret the veil as a type of social border limiting the woman's rights and fostering discrimination. In 2011 France banned wearing veils and other face coverings in public spaces. Other European governments have proposed similar bans. These domestic debates quickly jumped their nominal jurisdictions as both sides utilized new information technologies to influence global public opinion.

Other recent trends involve attempts to export particular forms of western feminism to developing countries. While proponents may portray these efforts as extending to women universalistic notions of human rights or embracing the neo-interventionist ideals of crossing international borders, opponents may note that feminism has different connotations in nonwestern cultures or even interpret this as simply the most recent example of western cultural imperialism. The field of international microfinance, for example, is rapidly becoming one of the most prevalent sources of social change in developing regions. Nongovernmental organizations (NGOs), like the Grameen Bank, FINCA International, World Vision, or Women's World Banking, specifically target female entrepreneurs to empower women across a variety of sociocultural contexts. The core principle is not simply the redistribution of power between the genders but the advancement of a particular agenda that transcends the filtering capacity of borders and state sovereignty. Such advancement inherently crosses social boundaries and requires broad-based social adjustments.

The Taliban's efforts to cordon off Afghanistan during the 1990s offer a clear example of state borders employed as tools to prohibit foreign cultural influences. Within Taliban-dominated Afghanistan, a wide variety of previously lawful activities were forbidden under a skewed interpretation of Sharia law combined with local tribal traditions. The prohibitions included: all pork products, products containing human hair, satellite dishes, cinematography, most musical instruments and audio equipment, pool tables, chess, masks, alcohol, cassette tapes, computers, VCRs, television, sexually evocative material, wine, lobsters, dancing, kites, nail polish, firecrackers, statues, and sewing catalogs. Employment, education, health care, and sports for women were drastically limited. For their part, men had to grow beards of a certain length, keep their head hair short, and don head coverings. Few recognized the role borders played in facilitating these extreme cultural reforms. Like Stalin's efforts to lower an "Iron

Curtain" across Eastern Europe, the Taliban sought to construct a "mud-brick wall" around Afghanistan within which their version of cultural purity might be pursued. The result was a realm of oppression for both women and men, as well as a haven for extremist groups like Al Qaeda seeking to project a global agenda.

Although western feminism, microfinance, and many other trends aim toward the dissolution of social boundaries, borders remain common tools for those seeking to propagate various forms of differentiation or even discrimination. These boundary processes influence the formation of a variety of substate borders, including voting districts, census tracts, and school zones. Such subdivisions of sovereign territory vary from state to state, but each allocates some degree of power to their respective governments. The borders of local, municipal, provincial, or federal entities mark off spaces of responsibility and levels of jurisdiction. Linkages within these hierarchies of places shape networks of human commitments, capacities, and strategies. They structure modern societies and, some would argue, constitute the basic building blocks of contemporary democratic practices. Indeed, it is unclear if democracy on any meaningful scale can form and function in the absence of discreet territorial entities.

The creation of legislative districts vividly illustrates the territorial basis for popular representation. Indeed, most democratic systems feature substate borders designed to ensure equal territorial, as well as demographic, representation. Although these procedures are highly vulnerable to deliberate manipulation for the purpose of influencing elections (i.e., gerrymandering), they remain a pragmatic means for combating voter fraud. Indeed, the prospects of election manipulation would escalate without territorial-based controls, such as proof of citizenship and residence within a local municipality. This does not, however, assuage the risk of higher-level corruption or historically institutionalized inequalities that skew legitimate democratic representation of ethnic, racial, or economic groups. Nevertheless, the democratic process requires setting some scope for popular participation and representation. In other words, state, province, or district borders help set practical and manageable parameters for polling the electoral body, as well as defining the jurisdictional realm of the elected.

Clearly, democracy as a practice of governance has come to be inexorably bound with territorial sovereignty. The transition from religious-monarchial sovereignty to popular-territorial sovereignty was facilitated by the unity of the people (real or imagined) and their sanction (real or imagined) of the state to represent them. Following the American and French revolutions, territorial democracy and popular sovereignty grew in tandem. Nationalism reinforced the link between these ideals through constructed histories of struggle and social organization, which symbolically framed the unity of the people within the confines of a particular territory. The rapid proliferation of this framework over the past two centuries solidified the relationship between democracy and territory, as well as demarcating the borders between contemporary nation-states. These linkages have become so embedded that criticizing the notion of territorial sovereignty is commonly interpreted as challenging the rule of "the people." Standard political maps naturalize this relationship by portraying the world as a collection of discreet territorial units. Yet, this obscures the complexity of cross-border relationships and the daily practices of integration that pervade the contemporary international scene.

Supranationalism and regionalism

The territorial assumptions of the nation-state system are also being challenged by the development of supranational organizations. "Supranationalism" refers to the process of states transferring portions of their sovereignty to a larger quasi-federal entity. Some of these supranational organizations play highly visible roles in international diplomacy and trade, such as the United Nations and World Trade Organization. In some cases, supranational organizations seem to be gradually assuming the function of sovereign states—only at a larger scale. For some, the emergence of supranational organizations is leading to a form of cosmopolitan democracy in which global citizenship and universal rights replace the current world of territorial belonging and state sovereignty. Most proponents of this ideal envision a scaling up to the level of global democracy, although no one has been able to offer a viable strategy for achieving this goal. History provides numerous examples of alliances between states, but the evolution of Europe from the tinderbox of two world wars to the standard-bearer for supranational integration is of special significance to the changing role of borders.

Rooted in the desire to stave off geopolitical tensions, enhance economic cooperation, and address cross-border problems like pollution and crime, the European Economic Community evolved over time into the European Union. This reinvention of the European political ideal seeks to make the premise of territorial sovereignty increasingly irrelevant for the expanding number of member states. The supranational government based in Brussels has orchestrated successive treaties between member states, which facilitate freedom of movement for goods, services, capital, and people. The adoption of the Euro common currency, visa-free travel within the European Union, standardized passports for external trips, and the ability to vacation, study, and work outside one's state of birth seems to suggest bordered nation-states will disappear or increasingly function as substate units with a larger federal structure.

Ironically, increased permeability of borders within the European Union has been accompanied by a hardening of the borders between EU and non-EU states. Today, EU states sharing a border with non-EU states are required to demonstrate efficient policing of their external borders. The notion of "Fortress Europe" has taken hold in the minds of many, as the freedom of movement within the European Union requires increased enforcement along external borders to check crime and illegal migration. Beginning in 2010, the continuing European debt and currency crisis has also eroded popular support for regional integration. Taken together, these trends serve as potential catalysts for reviving nationalist sentiments, which could in turn revive territorialities across the Continent.

It should be noted, however, that fiscal crisis, migrants, and crime are not the only catalysts for new bordering trends in Europe. The European Union has also embraced the principle of subsidiarity, which holds that governmental responsibilities should be handled at the lowest possible level of government. This has helped fuel a process of political decentralization as states such as Italy, Spain, France, and Belgium have gradually moved toward adopting more federal systems of governance since the 1970s. New regionalisms are also evident as provincial units or groups of units discover that the advantages of cross-border linkages and greater autonomy may outweigh the benefits of interaction with other areas of their nominal state. Local leaders seek to strike a complex balance between governance and external cooperation in a form of neoregionalism that enables the expression of local voices and maintenance of regional identity.

Indeed, the European Union's success in eliminating so many barriers between member states may ironically increase independence movements. It now appears increasingly feasible for smaller regions to gain independence while still retaining the economic and security advantages of EU membership. The Padania independence movement in northern Italy, with its clear sense of distinction from southern Italian culture, is a prime example. In other places, such as Scotland, Flanders, Catalonia, and Corsica, the reemergence of older regional frameworks combines with strong ethnic identities to challenge the territorial integrity of their respective states.

While the European Union is one of the most visible and developed supranational organizations, many others exist and provide important forums for cross-border relations. Many of these institutions focus on economic issues and trade, such as the Asia-Pacific Economic Cooperation or the Eurasian Economic Community. Others focus on security concerns and military cooperation, such as the Collective Security Treaty Organization or the North Atlantic Treaty Organization. Still others, like the Africa Union or Union of South American Nations, aim for deeper and broader cooperation modeled on the European Union. These organizations have generally been most effective in areas concerning economic cooperation, especially liberalization of trade. It is uncertain if lasting cooperation or greater integration will emerge on other issues.

Despite a general media focus on economics, much of the popular support for supranationalism around the world derives from social welfare issues, human rights, and environmental concerns. The early stages of European integration drew impetus from efforts to combat acid rain, polluted rivers, and industrial emissions. Today, substate and nonstate actors increasingly stand at the forefront of environmental causes and proclaim the simple truth that pollution, natural hazards, diseases, and moral concerns do not heed the barrier or filter functions of political borders.

Environmental concerns and borders

Though borders are human constructs, spatial concentrations of particular species and phenomena are certainly present in nature. Yet the dividing line between them is rarely, if ever, clear and absolute. Some animals mark their territory with scent and actually police their claimed areas, but only humans seek to impose permanent lines to mark off territories of ownership, access, and belonging. Nature, on the other hand, pays little heed to political borders and thus functions as the most pervasive of border crossers. The intrinsic trans-state character of natural phenomena constitutes an additional realm evidencing divergent roles of territoriality and sovereignty in the contemporary global system.

Perhaps in recognition of nature's indifference to political borders, governments are increasingly using borders to preserve or protect nature from human intrusion. Nature preserves and national parks represent efforts to safeguard natural settings by creating borders that exclude certain human activities. Established in 1872, Yellowstone National Park in the western United States is generally recognized as the first modern national park. Many other states have since established their own park programs. While ostensibly positive, these efforts often work in tandem with parochial tendencies to nationalize and politicize state territory. Yet even the most rigid of borders are regularly traversed by seasonal migrations of animals or by seeds and insects dispersed through wind and water. As a result, states increasingly recognize the advantages of cooperative efforts for environmental stewardship. The formation of the Kgalagadi Transfrontier Park across the borders of South Africa and Botswana, and La Amistad International Park between Costa Rica and Panama, are two examples of this emerging trend.

Unfortunately, borders can also hinder conservation efforts since states with lax or nonexistent environmental regulations are still recognized as sovereign and therefore entitled to manage their territory free from outside interference. In response, some international organizations, like the World Wildlife Fund and other environmental NGOs, have tried to intervene in support of creating parks or preserves for particular endangered species. By seeking to impose their own preferences across state borders, these NGOs may conflict with local traditions or the interests of local governments. The new geopolitical reality of interventionism appears to be gaining acceptance in the environmental realm, just as it is within the human rights and security realms. There are currently movements to institutionalize global standards governing fish and wildlife harvests, bio-reserves, water pollution, carbon emissions, hazardous waste disposal, and the testing of nuclear weapons. These appear to be noble objectives, but each raises complex cross-border issues.

Trade agreements outlining the acceptability of certain species, for example, are explicit acts of border control that relate to environmental concerns. Beginning in 1996, the Mad Cow disease scare in Europe offered a clear example where fear of an outbreak was used in conjunction with state borders to protect markets for local beef producers. This is also evident in efforts by states to establish fishing and hunting regulations, categories of endangered or threatened species, and timber harvest limits and replanting requirements. These policies have legitimate purposes but may be manipulated to serve specific interest groups. They are therefore inherently political acts relating to borders.

Taking a broader view of borders and environmental well-being, it is clear that nature has its own spatial logic. As such, increasing human mobility and our capacity to transfer species across space constitutes a highly problematic driver of environmental change. Realization of this fact has driven the issue to the fore of domestic politics and international relations. The infamous introduction of the rabbit and the cane toad into Australia disrupted regional ecosystems. Similar problems have arisen from invasive species in the United States, such as kudzu plants, Japanese knotweed, zebra mussels, or Asian carp. These invasive species have triggered extensive containment efforts and restrictions on the transportation of species. Unfortunately, these efforts have been largely unsuccessful. On a more benign note, the breeding of pandas in zoos in both the United States and China from the 1970s served as launching points for higher-level diplomatic relations. These types of conservation efforts can be integral tools for promoting cross-border cooperation, especially concerning the management of transboundary natural resources.

Borders can also complicate the management of natural resources, especially transborder water resources. The Colorado River, for example, originates in the United States but flows downstream to Mexico, yet nearly all of the water is diverted for use before the river reaches Mexico. The major river systems in Central Asia span several states, but these states have competing priorities for utilizing the water. The upstream states of Kyrgyzstan and Tajikistan want to release reservoir water through hydroelectric dams in the winter to generate heat and power. The downstream states of Uzbekistan and Turkmenistan want water released in the summer for irrigation. Similar tensions persist between Israel, Syria, Jordan, and the West Bank over usage of the Jordan River. These cases raise questions as to ethics, natural resource usage rights, and territorial sovereignty.

Health and borders

By defining the limits of state sovereignty and substate jurisdiction, borders define spaces marked by profound differences in standards of living and quality of life. Indeed, rights to employment, housing, movement, health care, education, self-expression, sexual orientation, and even family size vary significantly from country to country. Some countries have constructed elaborate systems to protect their citizens by regulating such things as guns, toys, appliances, food, medicines, tobacco, and alcohol. Other states do nothing or almost nothing in this regard.

Health is among the most variable aspects of quality of life, and international borders play a substantive role in shaping these differences. Almost every imaginable facet of health exhibits significant variation from state to state, including fertility and mortality rates, life expectancy, and access to professional health care. Life expectancy, for example, averages around eighty years old in many developed countries but only around forty-five years old in some of the poorest countries. Beyond differences in life expectancy, the actual causes of mortality also vary significantly from country to country. Mortality in developed countries is mostly attributable to a combination of lifestyle choices, like tobacco use or inactivity, and problems associated with old age, like heart disease or cancer. In contrast, major causes of mortality among populations in developing countries often include relatively basic medical problems, like respiratory infections or dysentery often worsened by malnutrition. Simply put, borders divide and create spaces and populations of vastly different medical problems and outcomes.

The notion of environmental injustice signifies uneven geographic exposure to adverse health effects from air and water pollution and a range of other environmental problems. Even on a local scale, exposure to environmental problems and their associated health risks is disproportionately concentrated in areas of low income, minority, or indigenous populations. For example, landfills are generally located near low-income neighborhoods. On a global scale, the harmful impacts of overfishing or logging are most severe in poorer, developing countries.

That said, numerous health issues also cross state borders. Disease is one obvious transborder issue. The very idea of an "epidemic" relates to the spread of disease across social and substate boundaries, while the term "pandemic" signifies that a particular illness has spread globally. Globalization and the increased mobility of various peoples around the world contribute to new vectors for disease diffusion. Cholera originated in South Asia, for example, but developed into a global health threat that has periodically wreaked havoc on populations around the world for centuries, most recently following the 2010 earthquake in Haiti.

HIV/AIDS has also diffused around the world. It is immune to international borders and social boundaries but is clearly more prevalent in impoverished regions. The disease continues to ravage many parts of Africa, where programs designed to slow the contagion are hampered by competing myths, long-standing social norms, and overall lower standards of health care. The borders of the states in which this disease is most prevalent have limited capacities to prevent its spread into or out of the country.

Territorial sovereignty, however, does succeed in inhibiting effective treatment of many diseases and other health problems around the world. The preference of some governments to allocate scarce funding to military purposes rather than health care for their respective populations is a simple example. A more complex example is the varied legality of certain medicines/drugs, abortion, or assisted suicide. Each of these is determined by state law. It is for this reason that borders remain an integral force for producing uneven health care practices and outcomes around the world. Numerous organizations, such as Doctors Without Borders or the World Health Organization, are border-crossing entities with the express purpose of redressing the disparities in human health and quality of life that have been institutionalized by international borders and the ideal of territorial sovereignty.

Such disparities are not, however, unique to the international sphere. Substate and social boundaries are equally influential in demarcating quality of life issues among domestic populations. Residential segregation based on socioeconomic status is common in many countries. This creates spatial patterns where socioeconomic differences between adjacent neighborhoods or communities both reflect and help create differences in the overall health and quality of life of respective populations. It may be fairly stated that political borders, both international and domestic, and social borders create differences in a range of quality of life issues, including health care, educational opportunities, exposure to pollutants, and availability of clean drinking water or sewage systems.

Ethics and borders

Given their role in marking out wide disparities concerning quality of life, borders raise numerous ethical issues. Indeed, the processes of border creation and maintenance are inseparable from moral assumptions and judgments. While societies since antiquity have possessed general codes of conduct, the actual prescribed and proscribed practices varied greatly. Traditionally, these moral concerns have derived from specific religious belief systems. Animist, eastern, and western theologies developed their own specific customs related to territory and borders ranging from procedures for property ownership within religious communities to relationships with lands governed by other religious groups. Colonialism provided a platform for the expansion of European notions of "proper" state conduct into global norms and conventions. Initial efforts during the nineteenth century, such as the First Geneva Convention, tended to focus on state responsibilities and acceptable conduct during wartime, such as the treatment of prisoners and civilians. These conventions were largely ineffectual as the development of "total war" during World War II led to massive civilian casualties, expulsions, and genocide.

These tragedies provided renewed impetus to develop and apply broad-reaching conventions of international law and human rights. In addition to affirming the centrality of state territorial sovereignty, the founding charter of the United Nations from 1945 also listed "promoting and encouraging respect for human rights and for fundamental freedoms for all" as one of its main purposes. In 1948 the United Nations issued the Universal Declaration of Human Rights (UDHR) to clarify the meaning of "human rights" and "fundamental freedoms." While the UDHR set out a concrete list of human rights, it was not a binding treaty and lacked monitoring and enforcement mechanisms. Instead, the observance of human rights was left to the discretion of the states, which often prioritized geopolitical or cultural agendas that ran counter to the lofty ideals expressed through the UDHR.

Many contemporary scholars and activists have since emphasized how the continuing primacy placed on territorial sovereignty and border enforcement allows the occurrence of various human rights violations. Indeed, states accused of human-rights abuses commonly assert the principles of state sovereignty and territorial integrity to block intervention. This impasse in contemporary international law has led some to argue that ethical and moral concerns, specifically protection of basic human rights, justify cross-border interventions. In this view, violations of sovereignty are less important than ensuring basic human rights are upheld irrespective of state sovereignty.

Events in Darfur, Sudan, since 2003 illustrate the inherent contradictions between state sovereignty and human rights. Some governments and human-rights NGOs have publicly characterized the actions of the Sudanese government against the residents of Darfur as a campaign of genocide. Therefore, foreign intervention was justified to protect civilians and uphold human rights. The Sudanese government and its allies have responded that these actions merely represent a sovereign state exercising the right to administer and police its territory free from outside interference. In this case and many others, judgments of what qualifies as an "international" crime against humanity versus the legitimate actions of a sovereign state are often secondary to other political and economic considerations.

The sporadic nature of human-rights enforcement has fueled efforts to develop universal standards of justice. After World War II, the victorious states created international courts to deal with German and Japanese war criminals. The United Nations later established special tribunals to investigate and prosecute those responsible for atrocities committed in Rwanda, Sierra Leone, and Yugoslavia during the 1990s. Proponents heralded these tribunals as progress toward applying broad-reaching ideals of justice, but these were relatively rare and ad hoc cases, which required broad international consensus among UN Security Council members, especially the five veto-holding permanent members.

The establishment of the International Criminal Court (ICC) based in The Hague, Netherlands, was intended to address these deficiencies by creating a permanent mechanism for indictment and trial of political leaders and military commanders responsible for war crimes and other human-rights violations.

Although the name of the institution suggests a broad purview and some advocated the court be endowed with universal jurisdiction, the ICC's actual power is highly limited. The court's jurisdiction covers only those states that ratify the treaty, but many, including the United States, Russia, China, and India, have refused to do so and thus remain outside the court's purview. Among states that have ratified the treaty, there is general agreement that acts of genocide, wars of aggression, and crimes against humanity fall under the jurisdiction of the ICC, but states often disagree when these specific terms apply. Tragically, governments remain reluctant to intervene in the internal affairs of other states even when faced with clear evidence of human-rights abuses.

The limited effectiveness of the ICC and human-rights protections in general have led some to conclude that the territorial foundations of the modern state system are simply incompatible with the establishment of global human rights and therefore immoral. This argument is most developed in regards to migration. Since states and NGOs are often prevented from crossing borders to address human-rights abuses or even the differences in quality-of-life issues noted earlier, many have argued that developed countries have a moral obligation to relax migration controls and allow all migrants to enter, not just those classified as political refugees. Indeed, Article 13 of the UDHR states: "Everyone has the right to leave any country." Based on this, some have argued that the right to *leave* any country should be extended to include the right to *enter* any country at will, since the right to leave a country is meaningless if other countries are unwilling to grant entry. This would grant all persons the right to migrate to places offering higher standards of living and morally compel all governments to accept such migrants. While such a proposal is extremely unlikely to be implemented, this discussion highlights increasing appreciation for the moral and ethical dimensions implicit in the varied processes of bordering.

10. De-mining work in Sri Lanka is part of a global movement organized by the International Campaign to Ban Landmines (ICBL).

NGOs are also increasingly active and effective cross-border players. This book has already noted the role of NGOs as advocates for media freedom, conflict resolution, economic development, environmental improvement, and women's rights. One of the most noteworthy NGOs is the International Campaign to Ban Landmines (ICBL). The ICBL is actually a coalition of hundreds of humanitarian NGOs cooperating to rid the world of landmines and cluster munitions. The coalition's efforts helped bring about the Ottawa Treaty banning certain landmines. More than 150 countries have now signed the treaty, although many leading powers, notably the United States, Russia, China, and India, have not. In recognition of its efforts, the NCBL was awarded the Nobel Peace Prize in 1997. The role of NGOs is unquestionably significant, but it is imperative to recognize that they lack the resources and authority to respond effectively to the disparities in human rights and standards of living in the contemporary state-based territorial system.

Epilogue: A very bordered future

This book has explored the material and symbolic dimensions of borders, the varied processes of "bordering," and their influence on people's daily lives. From local to global scales, borders are understood as both formal and informal institutions of spatial and social practice, as well as physical and symbolic markers of difference. Borders are products of the groups they bound, varying in impact and meaning according to individual circumstances. Rather than simple demarcations of places, borders are manifestations of power in a world marked by significant spatial differences in wealth, rights, mobility, and standards of living. It is for these reasons that borders are such an important topic of study.

Given the variable nature of borders, attempts to predict future trends have obvious limitations. Yet after reviewing the historical evolution of borders and the spectrum of contemporary border research, it is worthwhile to ponder what the future might bring, however speculative that might be. The Westphalian principles of territorial sovereignty and rigid borders were never absolute in practice, but they have been, and in large part remain, the dominant mode of thought concerning the political division of the world. Yet the international economic, political, environmental, and cultural linkages that characterize the processes of globalization seem to undermine this paradigm. For some, globalization is creating a "flat world," or a world of flows and networks gradually replacing a bounded world of places. Still others discount such dramatic predictions, instead arguing that traditional nation-states will remain the dominant actors domestically and internationally for the foreseeable future.

From our perspective, reality is much more complicated than either view acknowledges. We see today, and are likely to continue to witness, contradictory trends related to borders. Rather than remaining essentially static or being completely overturned, current events suggest the territorial assumptions and role of borders in the contemporary world are experiencing a period of transition and re-negotiation. Indeed, such periods have clear precedents. It was the emergence of new forms of socioeconomic organization that fueled the initial formation of the territorial nation-state model in western Europe. This model was exported through colonialism, gradually replacing the relatively flexible notions of territory and borders that generally predominated in other regions.

It appears likely that we are currently experiencing a similar transitional period as new modes of socioeconomic organization, activity, and identity emerge in the early twenty-first century. Yet, the result will almost certainly not lead to a complete de-territorialization and elimination of borders. Any de-territorialization will likely coincide with some type of broad-based re-territorialization. This will require corresponding processes of re-bordering, although they will differ in form, function, and scale from previous structures. Never immutable, territoriality remains a force in human action and organization, and as such the prospect of a perpetual de-territorialization leading to a borderless world is highly unlikely.

So what might this new territorialization look like? Given that borders are becoming more open to certain sets of people, institutions, or movements while simultaneously more closed to others, it is worth pondering if we are witnessing the emergence, or perhaps reemergence, of more flexible and volatile territorial networks like those that characterized medieval Europe and non-European areas before colonialism. This would imply modes of territoriality characterized by overlapping, contingent, and flexible hierarchies of political power. The growing size and influence of international corporations, supranational organizations, and NGOs, as well as substate groups like local governments, indigenous peoples, or diasporic communities, all challenge the notion of absolute state territorial sovereignty. Such trends, combined with state policies institutionalizing graduated sovereignty/citizenship and the proliferation of neoliberal economic spaces, could suggest the emergence of "neofeudal" sociopolitical networks in which certain classes and institutions garner broad privileges, while others face greater discrimination and regulation.

During this stage of shifting spatiality, borders embody multifaceted and contradictory roles. Perhaps more prevalent today than at any point in history, we are confronted with the challenge of reconciling the transportable and multiscalar nature of territory, belonging, and governance with the reality of our very bordered world. In short, borders still matter and will continue to play powerful roles in global political, economic, cultural, and environmental issues.

This Very Short Introduction has demonstrated that borders have been, are today, and will remain central components of the human experience for the foreseeable future. Reflecting this fact, a rich literary genre has emerged in recent years offering an array of novels, short stories, autobiographies, and dramas exploring the complex cultural beliefs, histories, and material circumstances affecting individuals and communities within border spaces. These works are particularly useful in augmenting academic discourses on undocumented border crossings, indigenous struggles for land rights, and deepening interdependencies among neighboring states. Through textured accounts of life on the border, these authors attempt to sensitize policymakers and broader publics to the realities of both their actions and inactions.

The bounding of space is an innate feature of human existence. Humans are essentially place-makers, creating order by utilizing our capacity to physically and mentally demarcate differences between social, political, cultural, economic, and environmental entities, processes, systems, and institutions. As a result, our world is crisscrossed by lines marking varied jurisdictions of authority, ownership, and opportunity. The field of border studies offers a rich venue for research into the changing nature of human social-spatial organization. It is imperative that we understand how borders are being reconsidered and reformulated in contemporary economic, environmental, cultural, and geopolitical practices if we are to improve our individual and collective capacities for action amid the dynamics of globalization.

Further reading

Much of the material discussed in this book is drawn from research articles, which appeared in a variety of academic journals across the social sciences and humanities. Some of the best journals for border research are *Geopolitics*, *International Studies Quarterly*, *Journal of Borderlands Studies*, *Political Geography*, and *Regional Studies*. The list below, obviously far from comprehensive, focuses on some of the most important recent books in border studies. Their bibliographies contain extensive citations to earlier works related to borders. Readers may also consult the Very Short Introductions on Empire, Geopolitics, Globalization, International Migration, International Relations, and Nationalism.

Agnew, John. *Globalization and Sovereignty*. Lanham, MD: Rowman and Littlefield, 2009.

Alcock, Susan, Terence N. D'Altroy, Kathleen D. Morrison, and Carla M. Sinopoli, eds. *Empires: Perspectives from Archaeology and History*. Cambridge: Cambridge University Press, 2001.

Anderson, Malcolm. *Frontiers: Territory and State Formation in the Modern World*. Cambridge: Polity, 1996.

Arts, Bas, Arnoud Lagendijk, and Henk van Houtum, eds. *The Disoriented State: Shifts in Governmentality, Territoriality and Governance*. Dordrecht: Springer 2009.

Bayly, C. A. *The Birth of the Modern World, 1780–1914: Global Connections and Comparisons*. Malden: Blackwell, 2004.

Brunet-Jailly, Emmanuel. *Borderlands: Comparing Border Security in North America and Europe*. Ottawa: University of Ottawa Press, 2007.

Buchanan, Allen, and Margaret Moore, eds. *States, Nations, and Borders: The Ethics of Making Boundaries*. Cambridge: Cambridge University Press, 2003.

Butlin, Robin. *Geographies of Empire: European Empires and Colonies c. 1880–1960*. Cambridge: Cambridge University Press, 2009.
Cerny, Philip. *Rethinking World Politics: A Theory of Transnational Neopluralism*. Oxford: Oxford University Press, 2010.
de Blij, Harm. *The Power of Place: Geography, Destiny, and Globalization's Rough Landscape*. Oxford: Oxford University Press, 2009.
Diener, Alexander C., and Joshua Hagen, eds. *Borderlines and Borderlands: Political Oddities at the Edge of the Nation State*. Lanham, MD: Rowman and Littlefield, 2010.
Elden, Stuart. *Terror and Territory: The Spatial Extent of Sovereignty*. Minneapolis: University of Minnesota Press, 2009.
Fichtelberg, Aaron. *Crime Without Borders: An Introduction to International Criminal Justice*. Upper Saddle River, NJ: Pearson, 2008.
Friedman, Thomas. *The World Is Flat: A Brief History of the Twenty-first Century*. New York: Farrar, Straus and Giroux, 2005.
Ganster, Paul, and David E. Lorey, eds. *Borders and Border Politics in a Globalizing World*. Lanham, MD: SR Books, 2005.
Gavrilis, George. *The Dynamics of Interstate Boundaries*. Cambridge: Cambridge University Press, 2008.
Goldsmith, Jack, and Tim Wu. *Who Controls the Internet? Illusions of a Borderless World*. New York: Oxford University Press, 2006.
Hastings, Donnan, and Thomas Wilson. *Borders: Frontiers of Identity, Nation and State*. Oxford: Berg, 1999.
Houtum, Henk van, Olivier Kramsch, and Wolfgang Zierhofer, eds. *B/Ordering Space*. Aldershot: Ashgate, 2005.
Kolers, Avery. *Land, Conflict, and Justice: A Political Theory of Territory*. Cambridge: Cambridge University Press, 2009.
Kütting, Gabriela. *The Global Political Economy of the Environment and Tourism*. Basingstoke: Palgrave Macmillan, 2010.
Levitt, Peggy. *The Transnational Villagers*. Berkeley: University of California Press, 2001.
Mann, Michael. *The Sources of Social Power: A History of Power from the Beginning to A.D. 1760*. Cambridge: University of Cambridge Press, 1986.
Migdal, Joel, ed. *Boundaries and Belonging: States and Societies in the Struggle to Shape Identities and Local Practices*. Cambridge: Cambridge University Press, 2004.
Morris, Ian, and Walter Scheidel, eds. *The Dynamics of Ancient Empires: State Power from Assyria to Byzantium*. Oxford: Oxford University Press, 2009.

Newman, David, ed. *Boundaries, Territory, and Postmodernity.* London: Frank Cass, 1999.

Nicol, Heather, and Ian Townsend-Gault, eds. *Holding the Line: Borders in a Global World.* Vancouver: UBC Press, 2005.

O'Leary, Brendan, Ian S. Lustick, and Thomas Callaghy, eds. *Right-sizing the State: The Politics of Moving Borders.* Oxford: Oxford University Press, 2001.

Ong, Aihwa. *Neoliberalism as Exception: Mutations in Citizenship and Sovereignty.* Durham, NC: Duke University Press, 2006.

Pécoud, Antoine, and Paul de Guchteneire, eds. *Migration without Borders: Essays on the Free Movement of People.* New York: Berghahn Books, 2009.

Popescu, Gabriel. *Bordering and Ordering the Twenty-first Century: Understanding Borders.* Lanham, MD: Rowman and Littlefield, 2012.

Power, Daniel, and Naomi Standen, eds. *Frontiers in Question: Eurasian Borderlands, 700–1700.* New York: St. Martin's Press, 1999.

Rajaram, Prem Kumar, and Carl Grundy-Warr, eds. *Borderscapes: Hidden Geographies and Politics at Territory's Edge.* Minneapolis: University of Minnesota Press, 2007.

Robertson, Craig. *The Passport in America: The History of a Document.* New York: Oxford University Press, 2010.

Sack, Robert. *Human Territoriality: Its Theory and History.* Cambridge: Cambridge University Press, 1986.

Sadowski-Smith, Claudia. *Border Fictions: Globalization, Empire, and Writing at the Boundaries of the United States.* Charlottesville: University of Virginia Press, 2008.

Salehyan, Idean. *Rebels Without Borders: Transnational Insurgencies in World Politics.* Ithaca, NY: Cornell University Press, 2009.

Sassen, Saskia. *Territory, Authority, Rights: From Medieval to Global Assemblages.* Princeton, NJ: Princeton University Press, 2006.

Spruyt, Hendrik. *The Sovereign State and Its Competitors: An Analysis of Systems Change.* Princeton, NJ: Princeton University Press, 1996.

Stein, Mark. *How the States Got Their Shapes.* New York: Smithsonian Books/Collins, 2008.

Torpey, John. *The Invention of the Passport: Surveillance, Citizenship and the State.* Cambridge: Cambridge University Press, 1999.

Trigger, Bruce. *Understanding Early Civilizations: A Comparative Study.* Cambridge: Cambridge University Press, 2003.

Wastl-Walter, Doris, ed. *Ashgate Research Companion to Border Studies*. Farnham: Ashgate, 2011.
Weaver, John. *The Great Land Rush and the Making of the Modern World, 1650–1900*. Montreal: McGill-Queen's University Press, 2003.
Winichakul, Thongchai. *Siam Mapped: A History of the Geo-Body of a Nation*. Honolulu: University of Hawaii Press, 1994.

Websites

Association for Borderlands Studies
www.absborderlands.org
Provides information about the association's publications, meetings, and other activities.

Borderbase
www.nicolette.dk/borderbase/index.php
Includes links to assorted news reports and pictures concerning international borders.

Centre for International Borders Research at Queen's University Belfast
www.qub.ac.uk/research-centres/CentreforInternationalBordersResearch/ABS-CIBRBibliography
Offers descriptions on the center's research, publications, and an extensive bibliography of border literature.

International Boundaries Research Unit at Durham University
www.dur.ac.uk/ibru
Includes current news database, conferences, and research focused on border conflict resolution.

Nijmegen Centre for Border Research at Radboud University Nijmegen
http://ncbr.ruhosting.nl
Contains information on the center's research, conferences, and seminars.

Index

A

B

C

D

E

V

W

Y

Z